Igniting The Cosmos Rocketry Wonders

Igniting The Cosmos Rocketry Wonders

Solomon Raj

Noble Publishing

CONTENTS

INDEX

Introduction

In the tremendous field of the universe, the quest for information and investigation has been a central quality of human life. From the earliest civilizations that looked upon the stars with marvel to the advanced time where state of the art innovation impels us past our natural limits, the excursion of disclosure has been constant. At the front of this odyssey stands the wonder of rocketry — a demonstration of mankind's immovable soul and limitless interest.

The tale of rocketry unfurls like an embroidery woven across the records of time, mixing the strings of creative mind, development, and desire. A story crosses ages, connecting old longs for trip with the stunning real factors of contemporary space investigation. This ensemble of human accomplishment reverberates through the ages, repeating the determined call to try to achieve the impossible and open the secrets that lie past our earthly lines.

The beginnings of rocketry are well established in the thoughts of visionaries and the mission for greatness. In old China, black powder, at first imagined for restorative and otherworldly purposes, coincidentally turned into the charge for early rocket tests. The exploring endeavors of people like Wan Hu, an unbelievable figure from Chinese fables who purportedly endeavored to arrive at the Moon utilizing a simple rocket seat, epitomize the basic longing to break liberated from natural limitations and climb to the sky.

As hundreds of years passed, the fire of interest consumed more splendid. The Renaissance started a reestablished interest in the normal world, and masterminds, for example, Konrad Kyeser imagined fantastical machines that could navigate the skies. However, it was only after the twentieth century that the combination of logical information and innovative ability touched off the genuine period of rocketry. Crafted by visionaries like Konstantin Tsiolkovsky, who figured out the hypothetical reason for rocket drive, and Robert H. Goddard, who changed hypothesis into reality with the send off of the world's most memorable fluid powered rocket in 1926, laid the basis for the enormous odyssey that anticipated.

The interchange of international pressures and mechanical advancement catalyzed rocket improvement during the mid-twentieth 100 years. The cauldron of The Second Great War saw the development of long range rockets, flagging a change in perspective in the use of rocket innovation. The V-2 rocket, created by Nazi Germany, exemplified the disastrous capability of these new weapons and established the groundwork for the post-war space race. As the Virus War pressure raised, the US and the Soviet Association became entangled in an intense contest to overcome the last boondocks.

The peak of this extraordinary contention unfurled with the noteworthy send off of Sputnik 1 on October 4, 1957. The metallic circle, conveying the heaviness of human desire, orbited the Earth, denoting the introduction of the space age.

The ensuing competition to accomplish critical achievements escalated, with Yuri Gagarin's spearheading circle around the Earth in 1961 and Neil Armstrong's notable moonwalk in 1969. These fantastic accomplishments extended the skylines of human investigation as well as exhibited the groundbreaking force of rocketry on a worldwide scale.

The coming of room investigation delivered another period of logical revelation, mechanical advancement, and global joint effort. Rockets advanced from instruments of fighting to vessels of harmony, conveying payloads of logical instruments, correspondence satellites, and, at last, human wayfarers. The Worldwide Space Station (ISS), an image of solidarity in the universe, turned into a demonstration of the cooperative capability of room investigation, with commitments from countries all over the planet uniting chasing information past Earth's environment.

The space transport period introduced another part in rocketry, offering a reusable and flexible stage for space missions. Notwithstanding, the innate dangers and difficulties related with human spaceflight turned out to be horrendously obvious with misfortunes like the Challenger and Columbia debacles, provoking a reconsideration of room investigation procedures. As the van program finished up, a shift towards automated missions and private-area drives unfurled, democratizing admittance to space and encouraging a different environment of investigation.

Lately, the confidential area has arisen as a main thrust in the proceeded with development of rocketry. Visionary business people like Elon Musk with SpaceX, Jeff Bezos with Blue Beginning, and others have reshaped the scene of room investigation. The improvement of reusable rocket innovation, aggressive designs for lunar and Martian colonization, and the expansion of business satellite send-offs have reimagined the conceivable outcomes of what can be accomplished past our home planet.

All the while, progressions in impetus frameworks, materials science, and man-made reasoning have additionally moved the abilities of rockets. Developments like particle impetus, which bridles the force of electrically charged particles for push, vow to change profound space investigation, empowering missions that were once considered improbable. These steps not just upgrade our ability to investigate the universe yet additionally make ready for a practical and getting through human presence past Earth.

The idea of lighting the universe epitomizes the exacting terminating of rocket motors as well as the start of human creative mind and the unending quest for information. As we stand at the limit of another period in space investigation, the universe coaxes with neglected domains and untold marvels. The Artemis program, with its obligation to return people to the Moon and drive humankind towards Mars, epitomizes the recharged energy and aspiration that describes contemporary rocketry.

In the journey to light the universe, challenges proliferate — specialized, calculated, and moral. The sensitive harmony among investigation and safeguarding, the capable utilization of divine assets, and the moral contemplations of potential extraterrestrial experiences require cautious route. Besides, as we expand our venture into the universe, the significance of global coordinated effort, moral stewardship, and maintainability turns out to be progressively foremost.

Lighting the universe isn't simply a question of arriving at heavenly bodies; it is tied in with fuel the flares of motivation, encouraging an aggregate feeling of marvel, and embracing the innate human soul of investigation. As rockets pierce the limits of our air, they convey with them the goals of previous eras, the fantasies of the people who thought for even a second to look heavenward and imagine a predetermination past the bounds of Earth.

This story is a festival of the trailblazers, the visionaries, and the designers who have moved mankind into the universe. It is an investigation of the past, present, and fate of rocketry — a tribute to the tireless quest for information, the dauntlessness to dream, and the unyielding soul that moves us towards the stars. Go along with us on this enormous excursion as we dig into the marvels of rocketry, where the start of motors messengers the climb of rockets as well as the rising of human expected in the tremendous, unknown breadth of the universe.

Chapter 1

Setting the Stage

In the immense region of the universe, on a little blue planet known as Earth, the stage was set for the many-sided dance of life. It was a world overflowing with variety, from the transcending mountains that kissed the sky to the profundities of the sea, where strange animals wandered in the dimness. The land was painted with tints of green, as woods extended their appendages towards the sky, and waterways wandered through valleys, cutting the scene with their delicate industriousness.

In the midst of this normal embroidery, humankind arose, an animal categories supplied with the remarkable capacity to think, make, and shape their environmental factors. The sunset painted the skyline with a kaleidoscope of varieties as civilizations rose and fell, abandoning a heritage carved in the chronicles of time. From the antiquated Mesopotamian support of progress to the great pyramids of Egypt, the human excursion unfurled across ages and landmasses.

As hundreds of years passed, domains thrived and wound down, each adding to the rich mosaic of mankind's set of experiences. The ascent of Greece delivered the introduction of a vote based system and reasoning, while the Roman Domain made a permanent imprint on administration and designing. The Silk Street turned into a course for social trade, connecting the East and the West in an embroidery of exchange and thoughts.

In the archaic period, palaces and church buildings specked the European scene, and knights wore protection to safeguard their domains. The Islamic Brilliant Age introduced a renaissance of learning, with researchers deciphering and protecting old texts. In the mean time, far toward the east, the traditions of China developed workmanship, science, and administration, molding the predetermination of a tremendous domain.

The Renaissance, with its restoration of workmanship and information, brought forth another time of edification. Scholars like Leonardo da Vinci and Galileo Galilei extended the limits of human grasping, testing laid out standards and starting a logical transformation. The Period of Investigation saw valiant mariners like Christopher Columbus and Ferdinand Magellan graph strange waters, revealing new grounds and societies.

With the beginning of the Modern Upset, the world saw a seismic shift. Production lines murmured with the rhythm of progress, and steam motors fueled trains that crossed mainlands. The charm of new boondocks enticed trailblazers to the American West, where they cut out estates and established the groundwork for a thriving country.

The twentieth century unfurled with remarkable speed and force. The world was pushed into the cauldron of two universal conflicts, where countries wrestled with the ghost of contention on a worldwide scale. The outcome of obliteration led to the Unified Countries, an encouraging sign for worldwide participation and harmony. The Virus War, an international battle between the US and the Soviet Association, cast a long shadow over the world, characterizing a period of philosophical strain and mechanical contention.

Headways in science and innovation sped up, introducing the advanced age. PCs shrank in size however filled in power, associating individuals across the globe through the Internet. The Data Age unfolded, changing the manner in which mankind conveyed, learned, and associated. The inflexible walk of progress brought about space investigation, with people wandering past the limits of Earth, trying the impossible.

However, in the midst of the wonders of progress, the planet gave testimony regarding natural difficulties. Environmental change, deforestation, and contamination cast a pall over the once-unblemished climate. The sensitive equilibrium of biological systems was disturbed, prompting a chorale of worries about the fate of the planet and the species that called it home.

In the 21st hundred years, the world ended up at a junction. Globalization interconnected economies, societies, and social orders, cultivating both extraordinary joint effort and unexpected difficulties. The ascent of virtual entertainment achieved another time of moment correspondence, molding public talk and affecting political scenes. The idea of public personality wrestled with the powers of multiculturalism, as social orders explored the intricacies of variety and incorporation.

As the worldwide local area confronted the phantom of pandemics and wellbeing emergencies, the strength of human soul radiated through. Propels in medication and biotechnology offered trust for the moderation of illnesses, while the world wrestled with the moral ramifications of logical forward leaps. The interconnectedness of countries became obvious as the reaction to worldwide difficulties required aggregate activity and participation.

In the domain of international relations, power elements went through an unpretentious change. Arising economies advocated for themselves on the world stage, testing customary authorities and reshaping the scene of impact. The tussle for assets, both regular and innovative, filled international strains, making a fragile harmony among collaboration and rivalry on the global stage.

The mission for supportable advancement acquired unmistakable quality as ecological worries became the dominant focal point. The direness to address environmental change provoked countries to rethink their strategies and embrace sustainable power sources. The Paris

Understanding turned into an image of worldwide obligation to defending the planet for people in the future, featuring the interconnectedness of environmental prosperity and human thriving.

Mechanical developments kept on rethinking the limits of probability. Man-made consciousness, AI, and quantum processing proclaimed another boondocks of capacities, promising both exceptional open doors and moral problems. The combination of science and innovation brought about bioengineering, obscuring the lines between the normal and the fake.

In the midst of these epochal movements, cultural designs encountered a transformation. The elements of work developed with the ascent of far off availability and the gig economy, testing conventional ideas of business and employer stability. Schooling changed with the appearance of web based learning stages, democratizing admittance to information however bringing up issues about the fate of conventional instructive foundations.

The texture of social connections was rewoven in the advanced age. Online entertainment stages became fields for self-articulation, activism, and the molding of popular assessment. The limits between the virtual and the genuine obscured as online networks prospered, impacting political developments and social patterns. The quick scattering of data presented difficulties of deception and the disintegration of confidence in laid out foundations.

As humankind plunged into the future, the investigation of room recaptured force. Mars turned into the following outskirts for human investigation, with missions expecting to lay out a human presence on the red planet. The fantasy of interstellar travel started minds, with dreams of far off exoplanets and the potential for extraterrestrial life dazzling the human soul.

In the passageways of strategy, the sensitive dance of worldwide relations proceeded. The Unified Countries wrestled with the intricacies of keeping up with harmony in a world set apart by provincial contentions and international competitions. The mission for demobilization and the anticipation of atomic expansion stayed squeezing worries, as the worldwide local area tried to explore the fragile harmony between public interests and the benefit of everyone.

Against the scenery of these great accounts, the singular accounts of individuals unfurled in bunch ways. The human experience, with its embroidery of euphoria and distress, wins and adversities, wove together the strings of individual lives into the more extensive account of mankind. Dreams were pursued, challenges were met, and the human soul persevered, strong even with difficulty.

In the domains of craftsmanship, writing, and culture, the human creative mind kept on taking off. Innovativeness had no limits as specialists investigated new types of articulation, pushing the limits of custom and testing cultural standards. Writing turned into a mirror mirroring the intricacies of the human condition, catching the climate of every period in the ink of words.

The performing expressions, from theater to music, gave a material to the outflow of human inclination and the investigation of all inclusive subjects. Innovation turned into a partner in the inventive flow, growing the conceivable outcomes of narrating and creative development. Computer generated reality and expanded reality opened new boondocks for vivid encounters, obscuring the lines between the crowd and the actual craftsmanship.

In the realm of sports, competitors stretched the boundaries of human potential, breaking records and challenging assumptions. The Olympic Games kept on being an image of global kinship, uniting competitors from different foundations to contend on the world stage. Sports turned into a stage for social activism, with competitors capitalizing on their leverage to advocate for equity and fairness.

As the phase of mankind's set of experiences unfurled, the shadows of difficulties and clashes persevered. Treacheries and imbalances cast a long shadow, provoking social orders to defy the devils of their past and take a stab at a more evenhanded future. Developments for social liberties, orientation balance, and civil rights picked up speed, testing settled in frameworks and reshaping the talk on common freedoms.

The interconnectedness of the world became obvious notwithstanding worldwide difficulties. The Coronavirus pandemic, a seismic occasion in the mid 21st hundred years, highlighted the delicacy of human life and the requirement for aggregate activity. Countries wrestled with the double goals of defending general wellbeing and relieving the monetary effect, featuring the complicated dance of interests on the worldwide stage.

Chasing after information, the wildernesses of science extended dramatically. Forward leaps in hereditary qualities, neuroscience, and man-made brainpower opened new vistas for figuring out the complexities of life and awareness. The moral ramifications of logical progressions turned out to be progressively perplexing, bringing up issues about the limits of human mediation in the normal request.

Religion and otherworldliness kept on assuming a significant part in the human experience. Confidence, in its horde structures, gave comfort and importance to people exploring the intricacies of presence. Interfaith exchange and the investigation of otherworldliness with regards to logical comprehension became roads for looking for shared belief in the midst of variety.

Notwithstanding existential inquiries, the investigation of reasoning and existentialism thrived. Scholars wrestled with the idea of cognizance, the importance of life, and the moral contemplations of human activities. The enduring journey for shrewdness and understanding reverberated through the passages of the scholarly world and the thoughts of people considering their spot in the universe.

As the account of mankind's set of experiences unfurled, the pages representing things to come stayed unwritten. The decisions made by people, networks, and countries would shape the direction of the human story. The difficulties of the present, from natural supportability to civil rights, requested insightful reflection and purposeful activity.

The stage was set, not only for the continuation of the human adventure however for a section yet to be composed. The material of presence anticipated the strokes of human undertaking, imagination, and strength. The complicated dance of life would proceed, with each step making a permanent imprint on the embroidery of time. The sunset cast its brilliant shine not too far off, flagging the commitment of another sunrise, where the human soul would keep on taking off, unbound by the limitations of the past.

1.1 Introduction to the world of rocketry and its historical significance

The domain of rocketry, a space where science, designing, and human creative mind combine, has made a permanent imprint on the embroidery of mankind's set of experiences. Rocketry, as a discipline, addresses the investigation of room and the saddling of strong powers to impel objects past the bounds of Earth. Its verifiable importance traverses hundreds of years, with establishes implanted in old civilizations and branches stretching out into the vast compasses of the universe.

The starting points of rocketry can be followed back to old China, where explosive was first created for restorative and catalytic purposes. The unintentional revelation of the dangerous properties of this blend made ready for the improvement of early rocket innovation.

In the ninth 100 years, Chinese creators started utilizing explosive filled cylinders to push straightforward rockets out of sight. These early gadgets were utilized for military purposes, making a simple starting point for the future development of rocketry.

As hundreds of years unfurled, information on rocketry spread along the Silk Street, arriving at the Center East and Europe. The Mongols, having gained the privileged insights of black powder from the Chinese, coordinated rocket innovation into their tactical techniques. By the thirteenth 100 years, rockets were utilized in clashes between different Islamic domains, displaying the exchange of information and innovation across societies.

In middle age Europe, the idea of rockets as military weapons built up some decent forward movement. The Skirmish of Crecy in 1346 saw the arrangement of rockets by the English against the French. In any case, the viable utilization of rockets in fighting confronted difficulties because of their eccentric directions and absence of accuracy.

The Renaissance carried with it a recovery of interest in technical disciplines, laying the foundation for the precise investigation of rocketry. Visionaries like Konrad Haas and Johann Schmidlap of Germany conceptualized and planned early rocket-fueled gadgets. Notwithstanding, it was in the seventeenth century that Sir Isaac Newton's laws of movement gave the hypothetical underpinnings to figuring out the physical science of rocket drive. Newton's Third Regulation, expressing that for each activity, there is an equivalent and inverse response, turned into a foundation of advanced science.

The nineteenth century saw critical headways in both hypothesis and application. Crafted by Russian researcher Konstantin Tsiolkovsky established the hypothetical starting point for space travel. Tsiolkovsky's rocket condition, formed in 1903, gave experiences into the connection between rocket speed and the speed of ousted gases, a basic part of rocket impetus. Tsiolkovsky's visionary thoughts, including the idea of multi-stage rockets, became instrumental in forming the eventual fate of room investigation.

At the same time, in the US, Robert H. Goddard, frequently hailed as the dad of current rocketry, led spearheading tests. In 1926, Goddard effectively sent off the world's most memorable fluid powered rocket in Reddish, Massachusetts. This undeniable a turning point, exhibiting the plausibility of controlled and supported rocket flight. Goddard's commitments reached out past down to earth accomplishments; his works and licenses laid the foundation for future improvements in rocket innovation.

The interwar period and the result of The Second Great War saw an intermingling of logical information, international rivalry, and military desires that impelled rocketry into another time. The Nazi system in

Germany, under the administration of Wernher von Braun, fostered the V-2 rocket, the world's most memorable long-range directed long range rocket.

The V-2, notwithstanding its staggering use as a weapon, addressed a jump forward in rocket innovation and established the groundwork for future space investigation.

With the finish of The Second Great War, the Partners looked to saddle the logical mastery of German scientific geniuses for serene purposes. Under Activity Paperclip, von Braun and other German architects were brought to the US, where they assumed urgent parts in the improvement of American rocketry. The V-2 rocket turned into the reason for the Redstone and Jupiter series of rockets, denoting the start of the US's excursion into space.

The Virus War contention between the US and the Soviet Association turned into an impetus for fast progressions in rocket innovation. The Soviet Association, prodded by the progress of the principal fake satellite, Sputnik 1, in 1957, accomplished one more achievement with the send off of Yuri Gagarin, the primary human in space, in 1961. The US answered with the Mercury and Gemini programs, laying the foundation for the Apollo program, which finished in the noteworthy Apollo 11 moon arriving in 1969.

The Apollo missions, especially the notable second when space traveler Neil Armstrong set foot on the lunar surface, caught the aggregate creative mind of humankind. The space race had not just accomplished its underlying objective of arriving on the moon yet had additionally displayed the capacities of human space investigation. The pictures and words communicated from the moon's surface reverberated as a demonstration of human inventiveness and assurance.

The post-Apollo period saw the rise of new players in the field of room investigation. The Space Transport program, started by NASA in 1981, introduced another period of reusable shuttle. The bus armada turned into a workhorse for sending satellites, directing logical investigations, and gathering the Global Space Station (ISS). The ISS, a cooperative exertion including numerous countries, turned into an image of worldwide participation in space investigation.

As the twentieth century progressed into the 21st 100 years, the scene of room investigation went through massive changes. Administrative space organizations ended up joined by privately owned businesses anxious to take part in the blossoming space industry. SpaceX, established by business person Elon Musk in 2002, turned into a pioneer in business space travel. The improvement of the Hawk and Mythical serpent shuttle checked achievements in the mission for reusable and practical space transportation.

All the while, progressions in scaling down and satellite innovation reformed the space business. Little satellites, including CubeSats, opened up additional opportunities for logical examination, Earth perception, and media communications. The democratization of room access turned into a reality as colleges, new businesses, and even people could plan and send off their own satellites into space.

The investigation of Mars picked up reestablished speed in the 21st hundred years. Mechanical missions, for example, NASA's Mars meanderers Soul, Opportunity, Interest, and Steadiness, gave significant information about the Martian surface and the chance of previous existence. The desire to send people to Mars turned into a main thrust, with projects like SpaceX's Starship planning to make interplanetary travel a reality in the next few decades.

The approach of the NewSpace period, described by expanded private-area inclusion, re-imagined the scene of rocketry. Organizations like Blue Beginning, established by Amazon's Jeff Bezos, entered the space race with aggressive objectives of making space venture out available to regular people. The idea of room the travel industry, once bound to the domain of sci-fi, accepted substantial structure as privately owned businesses created suborbital spaceflight capacities.

Past the bounds of our planetary group, the quest for exoplanets turned into a focal point of galactic examination. Telescopes like Kepler and TESS (Traveling Exoplanet Overview Satellite) recognized a great many exoplanets in the tenable zones of far off stars, touching off the creative mind with the chance of extraterrestrial life. The revelation of water on far off moons and planets further energized the mission for grasping the potential for life past Earth.

The job of global coordinated effort in space investigation kept on advancing. The European Space Organization (ESA), the Russian space office Roscosmos, the China Public Space Organization (CNSA), and other spacefaring countries participated in joint missions, logical examination, and the advancement of room foundation. The Artemis program, drove by NASA, meant to return people to the lunar surface and lay out maintainable lunar investigation fully intent on planning for future ran missions to Mars.

The meaning of rocketry stretches out past the items of common sense of room investigation. It typifies the human soul of investigation, development, and the quest for information. Rockets are not only vehicles to navigate the universe; they are images of human inventiveness pushing against the limits of the unexplored world. The investigation of room difficulties how we might interpret the universe, our place in it, and the potential for life past our home planet.

Additionally, rocketry has useful ramifications for life on The planet. Satellites in circle empower worldwide correspondence, weather conditions anticipating, route, and Earth perception for calamity the board. The improvement of cutting edge materials, drive frameworks, and life support advances for space make a trip frequently prompts developments with earthbound applications, impacting businesses going from medical care to broadcast communications.

1.2 Overview of key pioneers in rocket science

The historical backdrop of advanced science is joined with the narratives of amazing people who, through their vision, assurance, and logical sharpness, moved humankind into the domain of room investigation.

These trailblazers, traversing various times and social foundations, established the groundwork for the advanced field of rocketry, molding the direction of room investigation. Their commitments, frequently against the scenery of war, international competitions, and mechanical difficulties, have passed on a getting through heritage that keeps on moving ages of researchers and architects.

Perhaps of the earliest figure throughout the entire existence of rocketry is Konstantin Tsiolkovsky, a Russian researcher and visionary whose work in the late nineteenth and mid twentieth hundreds of years laid the hypothetical foundation for space travel. Brought into the world in 1857, Tsiolkovsky conquered actual difficulties, including deafness, to turn into a self-trained polymath. In 1903, he distributed a fundamental paper named "Investigation of Space through Rocket Gadgets," in which he verbalized the idea of utilizing multi-stage rockets filled by fluid forces to accomplish get away from speed. Tsiolkovsky's thoughts, however at first met with distrust, became central to the advancement of rocket innovation and space investigation.

Lined up with Tsiolkovsky's work, American researcher and specialist Robert H. Goddard autonomously made earth shattering commitments to rocketry. Brought into the world in 1882, Goddard is many times hailed as the dad of present day rocketry. In 1914, he got a patent for a rocket plan that used fluid fuel, an idea that obvious a takeoff from the strong fuel rockets of his time. In 1926, Goddard effectively sent off the world's most memorable fluid powered rocket in Reddish-brown, Massachusetts. His spearheading work laid the preparation for resulting advancements in rocket impetus and space investigation.

The crossing point of rocketry and fighting during The Second Great War presented one more key figure throughout the entire existence of advanced science - Wernher von Braun. Brought into the world in Germany in 1912, von Braun assumed a urgent part in the improvement of the V-2 rocket, the world's most memorable long-range directed long range rocket. The V-2, utilized by Nazi Germany in the last option phases of the conflict, addressed a critical mechanical jump and established the groundwork for future space investigation. After the conflict, von Braun, alongside other German researchers, was brought to the US under Activity Paperclip, where he turned into a focal figure in the improvement of American rocketry.

In the US, the post-The Second Great War time saw the development of a gathering of researchers and specialists known as the "Rocket Young men," drove by Theodore von Kármán. As the establishing overseer of the Fly Drive Research facility (JPL), von Kármán assumed an essential part in propelling rocket innovation and encouraging a culture of development. Under his administration, JPL created and sent off the principal effective American satellite, Voyager 1, in 1958. This accomplishment denoted the start of the US's dynamic association in space investigation.

The space race between the US and the Soviet Association during the Virus War time delivered one more notable figure in advanced science - Sergei Korolev. Frequently alluded to as the "Boss Architect," Korolev was the main thrust behind the Soviet space program. Brought into the world in 1907, his initial work in rocketry prompted the fruitful send off of Sputnik 1, the first fake satellite, in 1957. Korolev's administration additionally finished in Yuri Gagarin's noteworthy spaceflight in 1961, making him the main human to circle the Earth. Korolev's commitments were instrumental in laying out the Soviet Association as a spearheading force in space investigation.

As the US went for the gold, the Apollo program turned into a demonstration of the cooperative endeavors of numerous people, including German-conceived specialist and chief Kurt H. Debus. As the principal head of NASA's Kennedy Space Center, Debus administered the turn of events and send off of the Saturn V rockets that pushed the Apollo missions to the moon. His administration was necessary to the outcome of the Apollo program, which accomplished the noteworthy accomplishment of landing people on the lunar surface with Apollo 11 of every 1969.

The mid-twentieth century likewise saw the rise of a visionary business person and specialist - Elon Musk. While Musk may not be a conventional trailblazer in the verifiable sense, his effect on contemporary rocketry and space investigation is significant. Brought into the world in 1971, Musk established SpaceX (Space Investigation Advances Corp.) in 2002 with the aggressive objective of diminishing space transportation costs and in the long run colonizing Mars. Under Musk's administration, SpaceX fostered the Endlessly bird of prey Weighty rockets, alongside the Mythical beast shuttle, exhibiting the practicality of reusable rocket innovation and changing the financial aspects of room travel.

SpaceX's achievements incorporate the primary secretly subsidized, fluid powered rocket (Bird of prey 1) to arrive at circle in 2008 and the main secretly evolved shuttle (Winged serpent) to be recuperated effectively from circle in 2010. Musk's bold vision and obligation to propelling space investigation have revived the business, rousing another period of public and confidential cooperation in space.

One more compelling business visionary in the field of business space investigation is Jeff Bezos, the pioneer behind Blue Beginning. Laid out in 2000, Blue Beginning spotlights on creating reusable rocket advances and means to empower private human spaceflight. Bezos imagines a future where a huge number of individuals reside and work in space, and Blue Beginning's New Shepard and New Glenn rockets address ventures toward accomplishing that vision. Bezos, similar to Musk, plays had a urgent impact in reshaping the scene of present day rocketry.

The commitments of these vital trailblazers in advanced science stretch out past individual accomplishments. They address a continuum of development, each structure upon the establishments laid by ancestors and adding to the aggregate headway of rocket innovation.

From Tsiolkovsky's hypothetical bits of knowledge to Musk's aggressive endeavors, the excursion of rocketry mirrors humankind's unending journey to investigate the universe and push the limits of what is conceivable.

Notwithstanding these key figures, the scene of rocketry has been molded by incalculable specialists, researchers, and visionaries who have added to the field. The cooperative endeavors of global space organizations, privately owned businesses, and research establishments have moved advanced science higher than ever. As innovation proceeds to progress, and humankind looks toward the possibility of interplanetary investigation and then some, the tradition of these trailblazers fills in as a directing light, moving people in the future to try the impossible and investigate the secrets of the universe.

1.3 Setting the tone for the exploration of rocketry wonders in the cosmos

The investigation of rocketry, with its wondrous potential outcomes and infinite desires, entices humankind to set out on an excursion that rises above the limits of our home planet. The actual substance of rocketry is interlaced with the soul of investigation, advancement, and the unfaltering longing to unwind the secrets of the universe. As we set the vibe for digging into the marvels of rocketry, it is fundamental to perceive the significant effect this field has had on molding the course of mankind's set of experiences, from antiquated explores different avenues regarding black powder moved gadgets to the contemporary journeys for interplanetary travel and then some.

At its center, rocketry addresses a combination of science, designing, and creative mind, moving us toward the sky with a power that resists gravity itself. The inestimable artful dance of divine bodies, the confounding profundities of space, and the neglected boondocks of far off planets have long spellbound the human creative mind. Rocketry fills in as the mechanical key to open the doors to these grandiose miracles, permitting us to navigate the huge territories of the universe and witness the excellence and intricacy that lie past the limits of Earth.

The authentic embroidery of rocketry is woven with strings of resourcefulness and persistence. Antiquated civic establishments, uninformed about the heavenly wonders that looked for them, coincidentally found the fundamentals of rocket innovation. In antiquated China, the unplanned revelation of explosive set up for the earliest analyses with rocket drive. As basic cylinders loaded up with explosive were lighted, they took off out of sight, making an exhibition that obvious the beginning of mankind's relationship with controlled impetus.

These early examinations established the groundwork for the advancement of rocketry, as information on black powder and its unstable properties navigated social and geological limits. The Silk Street turned into a conductor for the trading of thoughts, prompting the reception of rocket innovation in the tactical methodologies of middle age Europe and the Islamic world.

Rockets, at first utilized as weapons of war, continuously changed into instruments of logical request and investigation.

The Renaissance time saw a recovery of interest in the innate sciences, giving the scholarly scenery to the efficient investigation of rocketry. Masterminds like Konrad Haas and Johann Schmidlap in Germany conceptualized early rocket-fueled gadgets, imagining the potential for controlled flight. Nonetheless, it was the notable work of Sir Isaac Newton in the seventeenth century that established the hypothetical starting point for figuring out the physical science of rocket drive. Newton's Third Law of Movement, which expresses that for each activity, there is an equivalent and inverse response, turned into a foundation of advanced science.

In the mid twentieth 100 years, the hypothetical basis laid by visionaries like Konstantin Tsiolkovsky and Robert H. Goddard made ready for functional headways in rocket innovation. Tsiolkovsky's idea of utilizing multi-stage rockets powered by fluid forces and Goddard's effective send off of the world's most memorable fluid energized rocket denoted a change in outlook. The stage was set for the extraordinary time of room investigation that would unfurl in the a long time to come.

The juncture of science, international relations, and battle during The Second Great War pushed rocketry into another direction. Wernher von Braun, a critical figure in the improvement of the V-2 rocket for Nazi Germany, would later add to the American space program. The post-war time frame saw the exchange of German rocket aptitude to the US under Activity Paperclip, forming the direction of the early American space program.

The Virus War competition between the US and the Soviet Association unfurled against the background of the space race. Sergei Korolev, the central originator of the Soviet space program, planned the send off of Sputnik 1, the first fake satellite, in 1957. The outcome of Sputnik was a pivotal occasion, igniting a contest that would finish in Yuri Gagarin turning into the main human to circle the Earth in 1961. The US, moved by the vision of President John F. Kennedy, answered with the Apollo program, prompting the memorable moon arriving in 1969.

The Apollo missions, with their famous symbolism of space travelers strolling on the lunar surface, caught the aggregate creative mind of humankind. Seeing the Earth from the moon, a delicate desert spring suspended in the immensity of room, highlighted the interconnectedness of our planet and the unfathomable potential for investigation. The investigation of the moon denoted the pinnacle of human accomplishment in space investigation, a demonstration of the exceptional capacities of rocketry.

As the space race unfurled, the cooperative endeavors of researchers, designers, and pilgrims became foremost. Theodore von Kármán, the establishing head of the Stream Drive Research center (JPL), assumed a significant part in propelling rocket innovation and cultivating a culture of development.

The Apollo program, under the administration of people like Kurt H. Debus, exhibited the cooperative ability of worldwide groups making progress toward a shared objective.

The space transport period, with its reusable rocket and the get together of the Global Space Station (ISS), introduced another part of room investigation. The ISS, an image of worldwide collaboration, turned into a microgravity research facility where space explorers led examinations and studies to propel how we might interpret life in space. The van armada, including the Challenger and Columbia orbiters, exhibited both the victories and difficulties intrinsic in space travel.

The turn of the 21st century saw the rise of privately owned businesses in the space business, with figures like Elon Musk and Jeff Bezos driving the charge. Musk's SpaceX, established in 2002, zeroed in on decreasing space transportation expenses and making interplanetary travel a reality. SpaceX accomplished huge achievements, including the advancement of the Endlessly bird of prey Weighty rockets, the Mythical serpent space apparatus, and the fruitful organization of the Starlink satellite heavenly body.

Jeff Bezos, through Blue Beginning, meant to empower private human spaceflight and encourage a future where a large number of individuals reside and work in space. Blue Beginning's New Shepard and New Glenn rockets represent a promise to reusable rocket innovations and a dream of room as a space for human settlement. The contribution of privately owned businesses has re-imagined the space business, encouraging development and extending the opportunities for space investigation.

The investigation of Mars, a planet with a getting through charm, turned into a point of convergence of mechanical missions in the 21st 100 years. Wanderers like Soul, Opportunity, Interest, and Diligence gave important experiences into the Martian scene and the potential for previous existence. The possibility of human missions to Mars, supported by visionaries like Elon Musk, addresses another boondocks in rocketry and interplanetary investigation.

In equal, the field of cosmology and astronomy has seen pivotal revelations connected with exoplanets, dark openings, and the vast microwave foundation. Telescopes and observatories like the Hubble Space Telescope, Kepler, and the James Webb Space Telescope have extended how we might interpret the universe, disclosing its endlessness and intricacy.

The democratization of room access has turned into a characterizing element of the contemporary space age. Little satellites, including CubeSats, empower colleges, new businesses, and even people to participate in space investigation and logical exploration. The idea of room the travel industry, once consigned to the domains of sci-fi, is currently turning into an unmistakable reality as privately owned businesses pursue offering suborbital and orbital spaceflights to regular folks.

As we set the vibe for the investigation of rocketry ponders in the universe, recognizing the double idea of these endeavors is basic. Rocketry has impelled humankind higher than ever as well as presented difficulties and moral contemplations. The flotsam and jetsam produced by space missions, the ecological effect of rocket dispatches, and the possible militarization of room are issues that request insightful reflection and dependable practices.

Additionally, the quest for extraterrestrial life and the moral ramifications of human colonization of different planets bring up complex issues about our part in the universe. The stories of investigation, development, and revelation should be combined with a principled consciousness of our obligations as stewards of both Earth and the heavenly domains we look to investigate.

The marvels of rocketry, ready at the convergence of logical request and human creative mind, set up for a future where the universe becomes an objective as well as an expansion of our shared perspective. The difficulties and wins of the past push us toward another time of grandiose investigation, where the secrets of far off systems, the potential for life past Earth, and the possibilities of interstellar travel entice us to wander further into the inestimable unexplored world.

All in all, the investigation of rocketry ponders in the universe encapsulates the unyielding human soul, pushing the limits of what is reachable and igniting a feeling of stunningness and miracle. From the earliest examinations with black powder in antiquated China to the contemporary endeavors of privately owned businesses like SpaceX and Blue Beginning, the direction of rocketry mirrors mankind's natural interest and head to investigate the immense scope of the universe. As we explore the vast expressive dance of divine bodies and navigate the unknown boondocks of room, the miracles of rocketry stand as a demonstration of our tenacious quest for information, our strength even with difficulties, and our getting through journey to disentangle the enormous embroidery that wraps us.

The domain of rocketry, with its hypnotizing ponders in the universe, coaxes mankind to leave on a heavenly excursion that rises above the limits of our earthly home. This investigation, powered by a union of logical inventiveness, innovative ability, and an aggregate human longing for disclosure, outlines a direction that stretches from old tests with black powder to the cutting edge longs for interplanetary travel and the colonization of far off universes. As we dive into the significant effect of rocketry on how we might interpret the universe, we experience a story woven with the strings of development, joint effort, and the tenacious quest for the unexplored world.

The verifiable genealogy of rocketry follows its underlying foundations to old China, where the fortunate revelation of explosive set up for the earliest tests in controlled drive. Straightforward cylinders loaded up with explosive, when lighted, took off into the skies, denoting the commencement of mankind's interest with the ability to defeat gravity.

These simple gadgets, with their searing paths, were harbingers of a future where rockets would turn into the vessels conveying us to the furthest reaches of the universe.

The transmission of information along the Silk Street worked with the dispersion of rocket innovation to different societies. Archaic Europe and the Islamic world saw the combination of rockets into military systems, changing them from instruments of battle into vehicles for logical investigation. The Renaissance period saw the recovery of interest in inherent sciences, giving the scholarly background to the methodical investigation of rocketry. Masterminds like Konrad Haas and Johann Schmidlap in Germany imagined the potential for controlled flight, establishing the reasonable starting points for the headways to come.

The seventeenth century delivered Sir Isaac Newton's laws of movement, giving the hypothetical supporting urgent to grasping the physical science of rocket drive. Newton's third regulation, expressing that for each activity, there is an equivalent and inverse response, turned into a crucial guideline in the field of advanced science. The logical preparation laid by Newton set up for the extraordinary time that unfurled in the twentieth 100 years.

In the mid twentieth 100 years, the hypothetical experiences of visionaries like Konstantin Tsiolkovsky and Robert H. Goddard introduced reasonable progressions in rocket innovation. Tsiolkovsky's vision of multi-stage rockets moved by fluid charges and Goddard's fruitful send off of the world's most memorable fluid energized rocket denoted a change in perspective. The beginning field of rocketry had progressed from hypothetical thoughts to substantial trial and error, getting the wheels under way for humankind's climb into space.

The interaction of science, international relations, and battle during The Second Great War moved rocketry into another direction. Wernher von Braun, a critical figure in the improvement of the V-2 rocket for Nazi Germany, would later contribute essentially to the American space program. The post-war time frame saw the exchange of German rocket mastery to the US under Activity Paperclip, forming the direction of the incipient American space investigation endeavors.

The Virus War contention between the US and the Soviet Association unfurled against the background of the space race, a rivalry that would impel rocketry to phenomenal levels. Sergei Korolev, the main fashioner of the Soviet space program, arranged the send off of Sputnik 1 of every 1957, denoting the coming of the space age. The progress of Sputnik was an extremely important occasion that lighted an intense contest, coming full circle in Yuri Gagarin's notable circle of the Earth in 1961. The US, prodded by the vision of President John F. Kennedy, answered with the Apollo program, eventually accomplishing the great accomplishment of landing people on the lunar surface with Apollo 11 of every 1969.

The Apollo missions, with their notable symbolism of space explorers strolling on the moon, caught the aggregate creative mind of humankind. Seeing the Earth from the lunar surface, a delicate desert garden suspended in the endlessness of room, highlighted the interconnectedness of our planet and the limitless potential for investigation. The investigation of the moon denoted the pinnacle of human accomplishment in space investigation, a demonstration of the remarkable capacities of rocketry.

In the midst of the victories and difficulties of the space race, co-operative endeavors among researchers, specialists, and travelers became central. Theodore von Kármán, the establishing head of the Stream Drive Lab (JPL), assumed an essential part in propelling rocket innovation and encouraging a culture of development. The Apollo program, under the administration of people like Kurt H. Debus, showed the co-operative ability of worldwide groups making progress toward a shared objective.

The space transport time, with its reusable space apparatus and the get together of the Global Space Station (ISS), introduced another section of room investigation. The ISS, an image of worldwide participation, turned into a microgravity research center where space travelers led trials and studies to propel how we might interpret life in space. The bus armada, including the Challenger and Columbia orbiters, displayed both the victories and difficulties innate in space travel.

The turn of the 21st century saw the rise of privately owned businesses in the space business, led by figures like Elon Musk and Jeff Bezos. Musk's SpaceX, established in 2002, zeroed in on diminishing space transportation expenses and making interplanetary travel a reality. SpaceX accomplished huge achievements, including the advancement of the Endlessly bird of prey Weighty rockets, the Winged serpent shuttle, and the effective sending of the Starlink satellite heavenly body.

Jeff Bezos, through Blue Beginning, planned to empower private human spaceflight and encourage a future where a large number of individuals reside and work in space. Blue Beginning's New Shepard and New Glenn rockets represent a pledge to reusable rocket innovations and a dream of room as a space for human settlement. The contribution of privately owned businesses has reclassified the space business, encouraging advancement and growing the opportunities for space investigation.

The investigation of Mars, a planet with a getting through charm, turned into a point of convergence of mechanical missions in the 21st hundred years. Wanderers like Soul, Opportunity, Interest, and Steadiness gave important bits of knowledge into the Martian scene and the potential for previous existence. The possibility of human missions to Mars, supported by visionaries like Elon Musk, addresses another boondocks in rocketry and interplanetary investigation.

In equal, the field of space science and astronomy has seen noteworthy disclosures connected with exoplanets, dark openings, and the vast microwave foundation. Telescopes and observatories like the Hubble Space Telescope, Kepler, and the James Webb Space Telescope have extended how we might interpret the universe, revealing its immeasurability and intricacy.

The democratization of room access has turned into a characterizing element of the contemporary space age. Little satellites, including CubeSats, empower colleges, new companies, and even people to participate in space investigation and logical examination. The idea of room the travel industry, once consigned to the domains of sci-fi, is currently turning into a substantial reality as privately owned businesses pursue offering suborbital and orbital spaceflights to regular people.

As we set the vibe for the investigation of rocketry ponders in the universe, recognizing the double idea of these endeavors is basic. Rocketry has pushed humankind higher than ever as well as presented difficulties and moral contemplations. The trash produced by space missions, the natural effect of rocket dispatches, and the expected militarization of room are issues that request insightful reflection and capable practices.

Besides, the quest for extraterrestrial life and the moral ramifications of human colonization of different planets bring up complex issues about our job in the universe. The stories of investigation, development, and revelation should be combined with an honest consciousness of our obligations as stewards of both Earth and the heavenly domains we look to investigate.

The miracles of rocketry, ready at the crossing point of logical request and human creative mind, set up for a future where the universe becomes an objective as well as an expansion of our shared perspective. The difficulties and wins of the past impel us toward another period of grandiose investigation, where the secrets of far off worlds, the potential for life past Earth, and the possibilities of interstellar travel coax us to wander further into the astronomical unexplored world.

All in all, the investigation of rocketry ponders in the universe typifies the dauntless human soul, pushing the limits of what is reachable and encouraging a feeling of stunningness and marvel. From the earliest tests with black powder in antiquated China to the contemporary endeavors of privately owned businesses like SpaceX and Blue Beginning, the direction of rocketry mirrors humankind's natural interest and head to investigate the immense territory of the universe. As we explore the enormous expressive dance of divine bodies and navigate the strange boondocks of room, the marvels of rocketry stand as a demonstration of our constant quest for information, our strength despite challenges, and our persevering through journey to unwind the infinite embroidery that encompasses us.

3 |

Chapter 2

The Birth of Rocketry

The introduction of rocketry can be followed back to antiquated times, where early developments established the groundworks for the rules that would ultimately impel humankind past the limits of Earth's air. The excursion from straightforward firecrackers to strong space-faring rockets is a demonstration of human interest, development, and the mission for investigation.

One of the earliest recorded occurrences of rocket-like gadgets traces all the way back to old China, around the ninth 100 years. Chinese innovators found that by filling bamboo tubes with explosive and fixing one end, they could make a drive framework. These early rockets were at first utilized for stately and military purposes, denoting the start of mankind's interest with tackling the dangerous force of explosive.

In the next hundreds of years, the information on rocketry spread to the Center East and Europe through exchange and military struggles. By the thirteenth hundred years, the Mongols had taken on Chinese rocket innovation and involved it in their successes. The spread of black powder related innovations assumed a significant part in forming the fate of rocket improvement.

As the Renaissance unfurled in Europe, logical reasoning and trial and error thrived. Visionaries like Konrad Haas and Johann Schmidlap in the sixteenth century conceptualized multi-stage rockets, a thought that would become key to space investigation hundreds of years after the fact. Nonetheless, progress in rocketry was slow, upset by restricted logical comprehension and an absence of useful applications.

It was only after the late nineteenth century that critical progressions happened. In 1888, Russian teacher Konstantin Tsiolkovsky, frequently viewed as the dad of astronautics, distributed a paper named "Investigation of Space through Rocket Gadgets." Tsiolkovsky's visionary work spread out the hypothetical structure for space travel, presenting ideas like multi-stage rockets, orbital mechanics, and the utilization of fluid charges.

While Tsiolkovsky was creating hypothetical standards in Russia, across the Atlantic, an American named Robert H. Goddard was directing viable examinations. In 1926, Goddard effectively sent off the world's most memorable fluid filled rocket in Reddish, Massachusetts. This noticeable a vital crossroads throughout the entire existence of rocketry, as it showed the possibility of controlled and coordinated rocket flight.

Goddard's work didn't get prompt acknowledgment, yet it laid the preparation for future turns of events. His licenses and compositions impacted later rocket pioneers, incorporating those associated with the V-2 rocket program drove by Wernher von Braun in Nazi Germany.

WWII turned into an impetus for fast headways in rocket innovation. The V-2, created by German architect Wernher von Braun, turned into the world's most memorable long-range directed long range rocket. In spite of its relationship with wartime obliteration, the V-2 addressed a mechanical jump forward and displayed the capability of rockets as weapons and devices for investigation.

As The Second Great War finished, the US and the Soviet Association went into another sort of contention, the Virus War. The space race arose as an emblematic and philosophical landmark between the two superpowers. In 1957, the Soviet Association accomplished a notable achievement by sending off Sputnik 1, the principal fake satellite, into space. This occasion denoted the start of the space age and escalated the space race between the US and the Soviet Association.

The US answered by laying out NASA in 1958 and setting out on the Mercury and Gemini programs. These drives intended to foster the innovation and aptitude required for human spaceflight. In 1961, Yuri Gagarin, a Soviet cosmonaut, turned into the primary human to circle the Earth, a critical accomplishment that further energized the space race.

On May 5, 1961, American space traveler Alan Shepard turned into the principal American in space during the Mercury-Redstone 3 mission. Shepard's suborbital flight was a critical stage toward additional aggressive objectives, like arriving at the Moon.

President John F. Kennedy's well known discourse in 1961 set up for quite possibly of the most notable crossroads throughout the entire existence of room investigation. He proclaimed the aggressive objective of handling an American space traveler on the Moon and returning them securely to Earth before the decade's end. This test, known as the Apollo program, turned into a main thrust behind headways in rocket innovation.

The Apollo program confronted various specialized difficulties, and the sad loss of the Apollo 1 team in a rocket fire featured the dangers related with space investigation. Notwithstanding, the assurance to arrive at the Moon won, and on July 20, 1969, the world saw Neil Armstrong and Buzz Aldrin become the main people to go to the lunar surface during the Apollo 11 mission.

The outcome of Apollo 11 was a demonstration of the cooperative endeavors of researchers, specialists, and space travelers who pushed the limits of human accomplishment. The space race had not just exhibited the capacity of rockets for space investigation yet additionally prodded progressions in different logical and mechanical fields.

In the next many years, space investigation kept on developing. The Space Transport program, started in 1981, addressed another time of reusable space vehicles. The buses assumed a urgent part in sending satellites, directing logical exploration, and gathering the Worldwide Space Station (ISS).

The ISS, a worldwide cooperative task including space organizations from the US, Russia, Europe, Japan, and Canada, turned into an image of global participation in space investigation. Development of the ISS started in 1998, and it has since filled in as a microgravity lab for logical examination and tests.

While government space organizations assumed a main part in the early long periods of room investigation, the last option a piece of the twentieth century saw the rise of privately owned businesses entering the space business. Business people like Elon Musk, with SpaceX, and Jeff Bezos, with Blue Beginning, meant to lessen the expense of room travel and make it more open.

SpaceX, established in 2002, accomplished a few achievements in the 21st 100 years, including the improvement of the Hawk 1, Bird of prey 9, and Bird of prey Weighty rockets. The organization's aggressive objectives incorporate decreasing space transportation costs, empowering the colonization of Mars, and making humankind a multi-planetary animal categories.

Progressions in rocketry were not restricted to conventional substance rockets. Ideas like particle drive and atomic warm impetus acquired consideration for their capability to alter space travel. Particle impetus frameworks, which utilize electrically charged particles for drive, offer higher effectiveness and efficiency for long-span missions.

The investigation of other heavenly bodies additionally kept on dazzling the creative mind of researchers and space lovers. Mechanical missions to Mars, for example, the Mars wanderers Soul, Opportunity, and Interest, gave significant information about the Red Planet's geography and history. The quest for extraterrestrial life and the comprehension of the planetary group's beginnings stayed key goals for space investigation.

As innovation progressed, the possibility of business space the travel industry turned into a reality. Organizations like Virgin Cosmic and Blue Beginning intended to give suborbital trips to regular folks, permitting private people to encounter a couple of moments of weightlessness and see the curve of the Earth.

The headway of rocketry likewise assumed a basic part in the improvement of satellite innovation, which became fundamental to correspondence, route, weather conditions checking, and Earth perception. Satellites in geostationary circle, for example, those utilized for broadcast communications, stay fixed comparative with Earth, giving persistent inclusion to explicit locales.

Lately, interest in returning people to the Moon and investigating Mars has reignited. NASA's Artemis program plans to land the principal lady and the following man on the lunar surface, preparing for economical lunar investigation. All the while, plans for manned missions to Mars are being investigated, with an accentuation on fostering the important advances for long-length space travel and life support.

The development of rocketry has been set apart by a persistent pattern of advancement, trial and error, and accomplishment. From the old Chinese innovators to the Apollo Moon arrivals and the ongoing period of private space investigation, rockets have impelled mankind into the universe, opening new wildernesses for revelation and understanding.

Looking forward, the eventual fate of rocketry holds commitment and difficulties. Headways in impetus frameworks, materials science, and man-made consciousness are supposed to shape the up and coming age of room investigation. Ideas like atomic warm drive and interplanetary rocket could empower quicker and more productive travel inside our planetary group.

Global coordinated effort stays essential, with drives like the Artemis Accords trying to lay out rules for tranquil and dependable investigation of space. As humankind proceeds to investigate and extend its presence in the universe, the illustrations gained from the introduction of rocketry will without a doubt direct the way forward, rousing people in the future to try the impossible and investigate the unexplored world.

2.1 Delving into the early experiments and innovations in rocket development

Diving into the early tests and developments in rocket improvement uncovers a rich embroidery of human creativity and interest that traverses hundreds of years. The excursion from basic firecrackers in old China to the improvement of refined rocket innovation in the cutting edge period is a demonstration of the persevering quest for information and the human craving to investigate the unexplored world.

The foundations of rocketry can be followed back to antiquated China, where, around the ninth hundred years, creators found the hazardous capability of black powder. This inadvertent revelation prompted the production of simple rocket-like gadgets. The Chinese found that by filling bamboo tubes with black powder and fixing one end, they could make a drive framework. At first utilized for stately and military purposes, these early rockets laid the preparation for future advancements in drive.

The spread of explosive innovation to the Center East and Europe worked with the trading of information and made way for additional turns of events. In the thirteenth 100 years, the Mongols embraced Chinese rocket innovation and integrated it into their tactical missions, acquainting these high level weapons with new districts. Notwithstanding these early progressions, the pragmatic utilizations of rocketry were restricted, and it remained generally bound to military and bubbly use.

As the Renaissance unfurled in Europe, logical reasoning and trial and error prospered. Visionaries of the sixteenth hundred years, like Konrad Haas and Johann Schmidlap, started to conceptualize multi-stage rockets, establishing the groundwork for ideas that would become critical in space investigation hundreds of years after the fact. In any case, progress in rocketry was slow, frustrated by an absence of comprehension of the hidden logical standards and restricted commonsense applications.

Hypothetical preparation for future advancements was laid by Russian researcher Konstantin Tsiolkovsky in 1888. Frequently viewed as the dad of astronautics, Tsiolkovsky distributed a paper named "Investigation of Space through Rocket Gadgets." In this work, he presented key ideas, for example, multi-stage rockets, orbital mechanics, and the utilization of fluid charges. Tsiolkovsky's visionary thoughts laid the hypothetical system for space travel and became instrumental in molding the eventual fate of rocketry.

While Tsiolkovsky was creating hypothetical standards in Russia, an American named Robert H. Goddard was directing viable tests that would bring the fantasy of room venture out nearer to the real world. In 1926, Goddard effectively sent off the world's most memorable fluid energized rocket in Reddish-brown, Massachusetts. This undeniable an essential crossroads throughout the entire existence of rocketry, exhibiting the possibility of controlled and coordinated rocket flight.

Goddard's work, however not promptly perceived, laid the preparation for future advancements in rocket innovation. His licenses and works affected later rocket pioneers, incorporating those associated with the V-2 rocket program drove by Wernher von Braun in Nazi Germany. The V-2 rocket, regardless of its relationship with wartime obliteration, addressed a jump forward in innovation and exhibited the capability of rockets as the two weapons and devices for investigation.

The finish of The Second Great War denoted a defining moment throughout the entire existence of rocketry. With the Partners accessing German rocket innovation and researchers, including Wernher von Braun, the establishments were laid for the space race that would characterize the Virus War time. The US and the Soviet Association arose as the primary players in a contest to show mechanical and philosophical predominance.

In 1957, the Soviet Association accomplished a memorable achievement by sending off Sputnik 1, the world's most memorable counterfeit satellite, into space. This occasion denoted the start of the space age and set off the space race between the superpowers. The US answered by laying out NASA in 1958 and leaving on the Mercury and Gemini programs, which planned to foster the innovation and skill required for human spaceflight.

On April 12, 1961, Yuri Gagarin, a Soviet cosmonaut, turned into the primary human to circle the Earth, a milestone accomplishment in space investigation. This achievement escalated the space race, and on May 5, 1961, American space explorer Alan Shepard turned into the primary American in space during the Mercury-Redstone 3 mission. Shepard's suborbital flight was a pivotal move toward additional aggressive objectives, like arriving at the Moon.

President John F. Kennedy's well known discourse on September 12, 1962, further stirred the American space exertion. In this discourse, Kennedy put forth a strong objective of handling an American space traveler on the Moon and securely returning them to Earth before the decade's end. This aggressive test, known as the Apollo program, turned into a main impetus behind headways in rocket innovation.

The Apollo program confronted various specialized difficulties, and the sad loss of the Apollo 1 group in a rocket fire highlighted the dangers related with space investigation. In any case, the assurance to accomplish Kennedy's objective won, and on July 20, 1969, the world watched in

stunningness as Neil Armstrong and Buzz Aldrin turned into the primary people to go to the lunar surface during the Apollo 11 mission.

The outcome of Apollo 11 was a victory of human accomplishment, exhibiting the cooperative endeavors of researchers, specialists, and space travelers. The space race not just exhibited the ability of rockets for space investigation yet additionally prodded progressions in different logical and mechanical fields. The Apollo program's heritage remembers developments for materials science, PC innovation, and frameworks designing.

In the many years that followed, space investigation kept on developing. The Space Transport program, started in 1981, addressed another period of reusable space vehicles. The buses assumed a vital part in sending satellites, leading logical exploration, and gathering the Worldwide Space Station (ISS). The ISS, a global cooperative venture including space organizations from the US, Russia, Europe, Japan, and Canada, turned into an image of worldwide collaboration in space investigation.

While government space organizations assumed a main part in the early long periods of room investigation, the last option a piece of the twentieth century saw the rise of privately owned businesses entering the space business. Business people like Elon Musk, with SpaceX, and Jeff Bezos, with Blue Beginning, intended to diminish the expense of room travel and make it more open.

SpaceX, established in 2002, accomplished a few achievements in the 21st 100 years, including the improvement of the Bird of prey 1, Hawk 9, and Hawk Weighty rockets. The organization's aggressive objectives incorporate decreasing space transportation costs, empowering the colonization of Mars, and making humankind a multi-planetary animal groups.

Progressions in rocketry were not restricted to customary compound rockets. Ideas like particle drive and atomic warm impetus acquired consideration for their capability to alter space travel. Particle drive frameworks, which utilize electrically charged particles for impetus, offer higher proficiency and efficiency for long-span missions.

The investigation of other heavenly bodies likewise kept on dazzling the creative mind of researchers and space aficionados. Mechanical missions to Mars, for example, the Mars meanderers Soul, Opportunity, and Interest, gave significant information about the Red Planet's geography and history. The quest for extraterrestrial life and the comprehension of the nearby planet group's beginnings stayed key targets for space investigation.

As innovation progressed, the possibility of business space the travel industry turned into a reality. Organizations like Virgin Cosmic and Blue Beginning intended to give suborbital trips to regular folks, permitting private people to encounter a couple of moments of weightlessness and see the curve of the Earth.

The progression of rocketry likewise assumed a basic part in the improvement of satellite innovation, which became vital to correspondence, route, weather conditions checking, and Earth perception. Satellites in geostationary circle, for example, those utilized for broadcast communications, stay fixed comparative with Earth, giving ceaseless inclusion to explicit areas.

Lately, interest in returning people to the Moon and investigating Mars has reignited. NASA's Artemis program means to land the principal lady and the following man on the lunar surface, making ready for maintainable lunar investigation. All the while, plans for maintained missions to Mars are being investigated, with an accentuation on fostering the fundamental advances for long-term space travel and life support.

The development of rocketry has been set apart by a nonstop pattern of advancement, trial and error, and accomplishment. From the antiquated Chinese innovators to the Apollo Moon arrivals and the ongoing time of private space investigation, rockets have moved mankind into the universe, opening new wildernesses for disclosure and understanding.

Looking forward, the eventual fate of rocketry holds commitment and difficulties. Progressions in drive frameworks, materials science, and computerized reasoning are supposed to shape the up and coming age of room investigation. Ideas like atomic warm impetus and interplanetary shuttle could empower quicker and more proficient travel inside our nearby planet group.

Global joint effort stays urgent, with drives like the Artemis Accords trying to lay out rules for serene and mindful investigation of space. As humankind proceeds to investigate and extend its presence in the universe, the illustrations gained from the early analyses and advancements in rocket improvement will without a doubt direct the way forward, rousing people in the future to try the impossible and investigate the unexplored world.

2.2 Profiles of visionaries like Konstantin Tsiolkovsky and Robert H. Goddard

Profiles of visionaries like Konstantin Tsiolkovsky and Robert H. Goddard offer a brief look into the personalities of people whose spearheading work established the groundwork for present day rocketry. These visionaries, isolated by geology and time, freely added to the hypothetical and functional parts of room investigation, molding the direction of mankind's excursion past the bounds of Earth.

Konstantin Tsiolkovsky: Father of Astronautics

Konstantin Tsiolkovsky, a Russian researcher and visionary, is in many cases hailed as the dad of astronautics. Brought into the world on September 17, 1857, in Izhevskoye, Russia, Tsiolkovsky's initial life was set apart by difficulties. He experienced hearing misfortune because of red fever, which essentially influenced his proper instruction. Regardless of these impediments, Tsiolkovsky's natural interest and love for learning drove him to seek after a great many interests.

Tsiolkovsky's interest with space and flight started in his childhood. In 1883, he distributed a paper named "Investigation of Space through Rocket Gadgets," a work that spread out the hypothetical structure for space travel. In this fundamental paper, Tsiolkovsky investigated the idea of utilizing multi-stage rockets energized by fluid fuels to accomplish space investigation. His visionary thoughts incorporated the idea of getting away from Earth's environment and using the vacuum of room for drive.

One of Tsiolkovsky's key commitments was the plan of the rocket condition, which depicts the material science of rocket impetus. This condition became essential to the comprehension of how rockets work and how they can accomplish the speed expected to arrive at space. Tsiolkovsky's bits of knowledge into the down to earth parts of rocketry showed a significant comprehension of the difficulties and conceivable outcomes related with space travel.

Tsiolkovsky's vision reached out past the specialized parts of rocketry. He anticipated the potential for human colonization of room and hypothesized about the cultural ramifications of room investigation. Tsiolkovsky accepted that the triumph of room could prompt the improvement of human culture and the amazing quality of natural constraints.

Notwithstanding the momentous idea of his thoughts, Tsiolkovsky's work got restricted acknowledgment during his lifetime. The confinement of Soviet Russia and the absence of a vigorous academic local area added to the generally calm affirmation of his commitments. It was exclusively in the last option part of the twentieth century that Tsiolkovsky's heritage acquired more extensive appreciation as his thoughts became basic to the space age.

Konstantin Tsiolkovsky's inheritance perseveres as a demonstration of the force of visionary reasoning. His conceptualization of room travel, drive frameworks, and the potential for human venture into the universe laid the foundation for resulting ages of researchers and specialists. Tsiolkovsky's impact reaches out past his particular commitments to rocketry, incorporating a more extensive vision of mankind's fate among the stars.

Robert H. Goddard: Trailblazer of Commonsense Rocketry

While Tsiolkovsky was figuring out hypothetical standards in Russia, Robert H. Goddard, an American physicist and designer, was leading viable trials that would carry rocketry from hypothesis to the real world. Brought into the world on October 5, 1882, in Worcester, Massachusetts, Goddard displayed an early interest in science and investigation. His childhood in a provincial climate cultivated a feeling of interest in the regular world and the conceivable outcomes of flight.

In 1914, Goddard got a patent for a rocket plan that consolidated a fluid powered motor. This undeniable a huge takeoff from prior strong fuel rockets and established the groundwork for the improvement of fluid powered rocket motors.

Goddard's patent illustrated a strategy for infusing fluid fuel into a burning chamber, showing a vital development that considered exact control of pushed.

Goddard's spearheading work didn't be ignored, however open and institutional acknowledgment was delayed to appear. His thoughts confronted doubt, and a few peers reprimanded his recommendations as unreasonable. Resolute, Goddard proceeded with his exploration and trials, accomplishing a progression of critical achievements.

On Walk 16, 1926, Goddard left a mark on the world by sending off the world's most memorable fluid powered rocket in Reddish, Massachusetts. The rocket, known as "Nell," arrived at an elevation of 41 feet and voyaged a distance of 184 feet. This effective flight showed the possibility of controlled and coordinated rocket flight, a key advancement that prepared for future improvements in rocket innovation.

Goddard's work pulled in the consideration of the U.S. military, and during The Second Great War, his examination added to headways in rocketry. Nonetheless, Goddard didn't live to see the full effect of his commitments. He died on August 10, 1945, preceding the finish of the conflict and the resulting speed increase of rocket advancement during the Virus War.

One of Goddard's enduring heritages is his obligation to the logical strategy and orderly trial and error. He carefully archived his work, keeping definite records of his analyses, computations, and experiences. Goddard's accentuation on innovative work as fundamental parts of advancement laid the foundation for the thorough methodology took on by people in the future of scientific geniuses.

Both Tsiolkovsky and Goddard, however working freely on inverse sides of the globe, shared a typical vision of room investigation and assumed crucial parts in the early improvement of rocketry. Tsiolkovsky's hypothetical commitments gave the scholarly system, while Goddard's down to earth tests exhibited the practicality of rocket-controlled flight.

The combination of these visionary thoughts and reasonable accomplishments laid the basis for the touchy development of rocket innovation during the mid-twentieth hundred years. The blend of hypothetical getting it and involved trial and error turned into a sign of resulting rocket improvement, impacting the researchers, designers, and space travelers who might proceed to shape the space age.

The traditions of Tsiolkovsky and Goddard reach out past their particular commitments to advanced science. They encapsulate the soul of investigation, advancement, and diligence that portrays humankind's journey to arrive at past the limits of Earth. The acknowledgment of their accomplishments fills in as a sign of the significance of visionary reasoning and the groundbreaking force of thoughts in forming the course of history.

As the investigation of room keeps on developing in the 21st 100 years, with aggressive designs for lunar missions, Mars investigation, and then some, the central work of visionaries like Tsiolkovsky and Goddard stays a persevering through wellspring of motivation. Their accounts advise us that the excursion into the universe isn't simply a logical and mechanical undertaking however a demonstration of the unlimited idea of human interest and the unyielding soul of investigation.

2.3 The foundational principles of rocket propulsion

The essential standards of rocket impetus support the whole field of advanced science, directing the plan, improvement, and activity of the vehicles that convey people and payloads into space. The standards overseeing rocket impetus are well established in the laws of physical science and thermodynamics, mirroring a fragile harmony between powers and responses that empower rockets to conquer Earth's gravitational force and navigate the limitlessness of space.

Newton's Third Law of Movement

At the core of rocket drive lies Newton's Third Law of Movement, which expresses that for each activity, there is an equivalent and inverse response. This basic standard is the foundation of rocketry. With regards to rocket drive, the activity is the ejection of mass, commonly as fumes gases, from the rocket's motors. The response is the subsequent forward pushed that drives the rocket the other way.

Rockets accomplish this push by removing high velocity gases through a rocket motor's spout. The ejection of mass produces a power as per Newton's Third Regulation, pushing the rocket forward. This standard is general and applies whether the rocket is sending off from Earth, moving in space, or arriving on another heavenly body.

Protection of Energy

The protection of energy is another basic rule administering rocket impetus. As indicated by this rule, the all out energy of a detached framework stays steady on the off chance that no outside powers follow up on it. With regards to a rocket, the framework incorporates both the rocket and the ousted exhaust gases.

At the point when a rocket ousts mass, it picks up speed the other way. The way to effective rocket drive is boosting the speed of the ousted mass. As per the rocket condition got from the preservation of force, the higher the speed of the ousted mass, the more prominent the subsequent change in speed (delta-v) of the actual rocket.

This relationship features the significance of accomplishing high exhaust speeds for effective rocketry. The journey for higher exhaust speeds has prompted the improvement of different impetus innovations, from conventional synthetic rockets to further developed ideas like particle drives and atomic warm impetus.

Push and Drive Effectiveness

Push, the power created by a rocket's motors, is a critical boundary in rocketry. It is the aftereffect of the mass stream pace of the ousted gases increased by their speed comparative with the rocket. Push drives the rocket against gravity and barometrical drag during climb.

The productivity of a rocket motor is much of the time evaluated by its particular motivation (ISP), which is a proportion of how successfully it changes over charge into push. Explicit motivation is the proportion of pushed to the weight stream pace of fuel and is communicated in short order. A higher explicit motivation shows a more effective motor, as it can create more push per unit mass of charge consumed.

Compound rockets, which rule current space investigation, accomplish push through the ignition of fuel. The substance responses discharge energy as hot gases, which are removed through a rocket spout to deliver push. The effectiveness of a compound rocket is impacted by the selection of charges, burning temperatures, and the plan of the motor.

Rocket Condition and Delta-V

The rocket condition, got from the preservation of force, measures the connection between the mass of the rocket, the mass of the removed charge, and the subsequent change in speed (delta-v). The condition is communicated as:

Δ

�

=

�

�

�

.

�

0

.

ln

(

�

0

�

�

)

Δv=I

sp

·g

0

·ln(

m

f

m

0

)

where:

Δ

�

Δv is the delta-v or change in speed,

�

�

�

I

sp

is the particular motivation of the rocket motor,

�

0

g

0

is the speed increase because of gravity at Earth's surface,

�

0

m

0

is the underlying mass of the rocket (counting force),

�

�

m

f

is the last mass of the rocket (after force has been ousted).

The rocket condition gives a central comprehension of the connection between the mass proportion of the rocket and the reachable change in speed. It underlines the remarkable idea of the relationship, featuring the difficulties of conveying adequate fuel to accomplish high delta-v prerequisites.

The delta-v idea is pivotal in mission arranging. Various periods of room missions, like arriving at circle, moving among circles, and leaving the Earth-Moon framework, require explicit delta-v qualities. Mission organizers utilize the rocket condition to decide the mass proportions expected for each stage and streamline charge use for effectiveness.

Kinds of Rocket Drive

Rocket drive frameworks come in different structures, each intended to address explicit mission prerequisites and difficulties. Compound rockets, which utilize substance responses to deliver fast exhaust gases, stay the most well-known impetus framework for sending off payloads into space. Fluid rocket motors, similar to those utilized on the Saturn V that conveyed space travelers to the Moon, consider exact control of pushed and can be closed down and restarted.

Strong rocket motors, with fuel in a strong state, are frequently utilized as sponsors for beginning push during send off. They give effortlessness and dependability however miss the mark on controllability of fluid rockets. Crossover rockets consolidate highlights of both fluid and strong impetus, with a fluid oxidizer and a strong fuel, offering benefits concerning security and control.

High level drive ideas are likewise being investigated for future space investigation. Particle drives utilize electric or electromagnetic fields to speed up particles, accomplishing extremely high unambiguous motivation and productivity over lengthy terms. Atomic warm impetus, an idea tracing all the way back to the mid-twentieth 100 years, includes utilizing an atomic reactor to warm fuel for push. While still in the calculated stage, these advancements address possible roads for more effective and aggressive space investigation.

Difficulties and Contemplations

While the essential standards of rocket impetus give the hypothetical premise to space travel, the pragmatic execution of these standards includes tending to various difficulties and contemplations. A portion of the key elements include:

Primary Uprightness: Rockets should endure outrageous powers during send off, including the extraordinary vibrations and streamlined pressures experienced during air rising. Underlying uprightness is vital to forestall horrendous disappointment.

Streamlined features: Rockets should be intended to limit streamlined haul during rising through Earth's climate. Smoothed out shapes and cautious thought of the rocket's direction assist with streamlining productivity.

Force Decision: The decision of charge fundamentally impacts rocket execution. Factors, for example, energy content, ignition security, and simplicity of taking care of assume a part in choosing reasonable charges for a given mission.

Heat The executives: The outrageous temperatures created during rocket drive require viable intensity the board. Warm assurance frameworks safeguard the rocket from serious intensity during reemergence into Earth's environment or other divine bodies.

Control Frameworks: Exact control of the rocket's direction and direction is fundamental for mission achievement. Response control frameworks, spinners, and direction frameworks guarantee that the rocket follows its planned way.

Payload Limit: The mass a rocket can convey to a particular circle, known as its payload limit, is a basic thought. Expanding payload limit frequently includes streamlining the rocket's plan and impetus effectiveness.

Cost Contemplations: The financial matters of room travel assume a critical part in the decision of drive frameworks. Accomplishing financially savvy admittance to space is a continuous test that impacts the plan and activity of send off vehicles.

Mission Arranging: Various missions require explicit delta-v capacities. Mission organizers should cautiously work out the delta-v necessities for each period of a mission and configuration rockets as needs be.

Tending to these difficulties includes a blend of designing mastery, mechanical development, and iterative plan processes. As rockets become more modern and space investigation targets become more aggressive, consistent upgrades in impetus innovation are fundamental.

The Fate of Rocket Impetus

The eventual fate of rocket impetus holds guarantee for additional progressions, driven by the quest for more aggressive space investigation objectives. Arising innovations and novel drive ideas are being investigated to improve the productivity, wellbeing, and manageability of room travel.

One area of dynamic exploration is the advancement of reusable rocket frameworks. Reusability means to lessen the expense of admittance to space by permitting rockets to be recuperated, revamped, and flown once more. Organizations like SpaceX have taken critical steps toward this path with the Hawk 9 and Bird of prey Weighty, showing the possibility of recuperating and reusing rocket parts.

Electric impetus, for example, particle drives, is acquiring consideration for giving productive drive overstretched mission durations potential. While flow particle drives are commonly utilized for profound space missions because of their low pushed, headways in electric impetus might prompt all the more remarkable frameworks reasonable for a more extensive scope of utilizations.

Atomic warm impetus, an idea investigated during the twentieth 100 years, is encountering restored revenue. The potential for atomic warm rockets to give high push and effectiveness makes them a possibility for manned missions to Mars and different objections past Earth.

Past Earth's circle, where sunlight based power is less viable, atomic electric drive is being considered for future profound space investigation. This idea includes involving an atomic reactor to produce power for particle drives, giving proficient impetus to mechanical and human missions to the external planets and then some.

Notwithstanding mechanical progressions, worldwide coordinated effort in space investigation is forming the fate of rocket impetus.

Cooperative endeavors, for example, those found in the development and activity of the Global Space Station (ISS), empower the pooling of assets and skill for additional aggressive missions.

As humankind focuses on getting back to the Moon, investigating Mars, and wandering farther into the universe, rocket drive will keep on advancing. The essential standards laid out by visionaries like Tsiolkovsky and Goddard will direct people in the future of researchers, designers, and space pioneers as they push the limits of what is conceivable.

The investigation of room stays a demonstration of human interest, inventiveness, and the aggregate drive to disentangle the secrets of the universe. With each send off into the universe, rockets convey payloads of logical instruments and space travelers as well as the fantasies and desires of an animal types headed to investigate the unexplored world.

4 |

Chapter 3

Rockets Reach for the Stars

In the huge region of the universe, humankind's steady interest has driven it to try to achieve the impossible. The excursion started with humble advances, grounded in the interest with the night sky that has dazzled personalities since days of yore. From the early legends and stories that wove stories of heavenly creatures to the logical requests that looked to unwind the secrets of the universe, the craving to investigate the universe has been a persevering through part of the human soul.

As the pages of history turned, so did the direction of mankind's yearnings. The coming of present day science and innovation denoted an essential second in the journey to disentangle the mysteries of the universe. The development of the telescope, credited to visionaries like Galileo Galilei, gave humankind a freshly discovered capacity to look past the bounds of the Earth. This innovative wonder uncovered a universe overflowing with cosmic systems, stars, and divine miracles, pushing the limits of human getting it.

Notwithstanding, the craving to investigate rose above simple perception. It appeared in fantasies about arriving at the universe, wandering past the gravitational hug of Earth to contact the actual texture of room itself. The idea of room travel turned into a tempting possibility, powered by crafted by visionaries, for example, Jules Verne, whose creative mind took off past the limitations of his time. Verne's "From the Earth to the Moon" charmed the personalities of perusers with its striking vision of a shot discharged from a huge cannon, conveying bold wayfarers to the moon.

As the twentieth century unfolded, the fantasies of room investigation started to appear into substantial plans. The hypothetical basis laid by visionaries tracked down unmistakable articulation as rocketry. The visionary scientific genius Konstantin Tsiolkovsky, frequently hailed as the dad of astronautics, enunciated the essential standards of rocket impetus. His momentous work established the groundwork for the mechanical jump that would drive mankind past the limits of Earth's environment.

The rise of rocketry as a practical method for space investigation matched with the wild occasions of the mid twentieth 100 years. The world saw the staggering force of rockets in fighting during The Second Great War. The V-2 rocket, created by Nazi Germany, addressed a lethal indication of rocket innovation. In any case, as the conflict reached a conclusion, exactly the same innovation that had fashioned obliteration turned into an encouraging sign for tranquil investigation.

In the result of The Second Great War, the US and the Soviet Association arose as superpowers secured in a Virus War contention. The international rivalry reached out past Earth's limits into the unknown domains of room. The Space Race was conceived, a wild contest to accomplish achievements in space investigation that would show mechanical ability and philosophical prevalence.

On October 4, 1957, the Soviet Association accomplished a notable achievement that resounded all over the planet. Sputnik 1, the primary fake satellite, circled the Earth, transmitting radio transmissions that repeated mankind's introduction to the universe. The blare signal blare of Sputnik's signs stamped an innovative victory as well as an emblematic second that rose above public lines. The time of room investigation had unfolded, and the world remained at the edge of another outskirts.

The US, decided not to be outperformed, answered with energy. On July 20, 1969, the Apollo 11 mission accomplished an accomplishment that had whenever been consigned to the domain of sci-fi. Space explorers Neil Armstrong and Buzz Aldrin set foot on the lunar surface, while Michael Collins circled above in the order module. The words articulated by Armstrong as he slid the stepping stool — "That is one little step for [a] man, one monster jump for humankind" — resounded across the globe, epitomizing the aggregate accomplishment of mankind.

The moon arrival was a perfection of long periods of logical and mechanical development. It exhibited the capacities of rocketry, route, life emotionally supportive networks, and the cooperative endeavors of innumerable people. The Apollo program showed that mankind had the ability to leave Earth and investigate other heavenly bodies. However, as the Apollo missions wound down, so did the enthusiasm for profound space investigation.

The 1970s and 1980s saw a change in center, with the improvement of room transport innovation and the foundation of low Earth circle tasks. While space investigation proceeded, the terrific vision of wandering past the moon blurred out of spotlight. The Space Transport turned into a workhorse, moving space travelers and payloads to and from low Earth circle, however the fantasies of interplanetary travel appeared to slip further from prompt reach.

Notwithstanding, the fire of room investigation won't ever smother. Visionaries, researchers, and designers kept on pushing the limits of what was conceivable. The fantasy about arriving at the stars continued, and as the 21st century unfurled, another time of room investigation started to come to fruition.

Privately owned businesses, once bound to the job of satellite send off suppliers, arose as central participants in the space race. SpaceX, established by Elon Musk in 2002, turned into a pioneer in rethinking the space business. Musk's vision reached out a long ways past the bounds of Earth, planning to make humankind a multiplanetary animal categories. The improvement of the Bird of prey and Starship rocket frameworks addressed a strong move toward accomplishing this brassy objective.

The development of rocket innovation was not restricted to the US. Different countries, perceiving the vital and logical significance of room investigation, increased their endeavors. China, specifically, took critical steps with its space program. The Chang'e missions to the moon, the Tiangong space station, and aggressive designs for Mars investigation displayed China's obligation to propelling its capacities in space.

As the world saw the resurgence of interest in space investigation, global coordinated efforts turned into a sign of the new time. The Worldwide Space Station (ISS), an image of participation among space-faring countries, circled the Earth, filling in as a microgravity research center and a demonstration of the potential for tranquil coordinated effort in space.

The revived excitement for space investigation was not bound to administrative offices and confidential undertakings. Established researchers assumed a pivotal part in molding the story of room investigation. Automated missions, furnished with cutting edge instruments, dug into the secrets of our planetary group and then some. Meanderers investigated the Martian surface, telescopes looked into far off systems, and tests wandered into the external ranges of our vast area.

The mission for tenable exoplanets turned into a point of convergence of galactic exploration. The disclosure of exoplanets inside the livable zones of far off stars energized the creative mind, raising the likelihood that other life-supporting universes could exist past our nearby planet group. The quest for extraterrestrial life, once consigned to the domain of sci-fi, turned into a real logical undertaking.

The longing to investigate the universe reached out past the bounds of our own world. The idea of interstellar travel, when thought about the stuff of imagination, acquired hypothetical establishing. Researchers and specialists considered the difficulties of arriving at other star frameworks, mulling over ideas, for example, twist drives and age sends that could convey people in the future of travelers on legendary excursions across the interstellar void.

The acknowledgment that the stars were not inaccessible places of light but rather objections inside the domain of human possible powered the improvement of aggressive tasks. Advancement Starshot, an idea supported by illuminating presences like Stephen Peddling, planned to send little, light-moved rocket to the Alpha Centauri star framework. The fantasy about arriving at one more star inside a human lifetime appeared to be implausible, yet the very thought denoted a change in perspective in how mankind saw its spot in the universe.

Mechanical progressions kept on moving the desires of room investigation. The improvement of particle drive, sun powered cruises, and high level life emotionally supportive networks opened additional opportunities for broadened space missions. Ideas once excused as sci-fi progressively turned into the focal point of serious logical request and designing advancement.

As humankind's range stretched out toward the stars, the moral and philosophical ramifications of room investigation came to the front. Inquiries regarding the protection of unblemished conditions on heavenly bodies, the potential for tainting different universes with earthbound life, and the moral contemplations of colonizing different planets requested cautious reflection. The possibility of experiencing extraterrestrial life, whether microbial or astute, brought up significant issues about our position in the enormous woven artwork.

The intermingling of man-made reasoning and space investigation further extended the skylines of plausibility. Independent space apparatus, directed by modern computer based intelligence frameworks, explored the universe with accuracy. AI calculations filtered through tremendous datasets from telescopes and tests, uncovering stowed away examples and opening the privileged insights of the universe.

The mission to arrive at the stars was not without its difficulties. The immense distances between heavenly bodies, the brutal states of room, and the restrictions forced by the laws of physical science introduced considerable obstructions.

However, the dauntless soul that pushed mankind from the caverns to the stars endured. The craving to investigate, to push limits, and to look for replies to the key inquiries of presence drove the tenacious quest for information and disclosure.

The excursion from the send off of Sputnik to the contemporary period of interstellar desires denoted a momentous section in the human story. Rockets, when images of obliteration, became vessels of investigation and revelation. The stars, when far off and unapproachable, became waypoints in mankind's vast excursion.

As we look toward the sky, the rockets that try the impossible represent mechanical ability as well as the persevering through human soul. The journey for information, the quest for understanding, and the longing to investigate the obscure keep on molding the direction of our species. Rockets, with their searing motors pushing them through the void, convey the goals of previous eras, present, and future.

In the fabulous embroidery of the universe, each rocket send off is a join, meshing together the tale of humankind's odyssey into the extraordinary unexplored world. The difficulties and wins, the mishaps and headways, all add to the rich story of our enormous excursion. Rockets try to achieve the impossible, not simply as vessels of metal and fuel, yet as expansions of the human soul arriving at past the constraints of what was once considered unimaginable.

What's to come holds commitment and vulnerability in equivalent measure. The visionaries of today, similar as their ancestors, wrestle with the intricacies of room investigation. The fantasies of interplanetary travel, of laying out settlements on Mars, of wandering past our nearby planet group, remain tantalizingly close yet slippery. The way ahead requires mechanical development as well as moral contemplations, global cooperation, and a profound comprehension of the significant effect our activities in space might have on the fate of our species and the actual universe.

As the rockets of the 21st century take off into the infinite chasm, they convey with them the deepest desires of innumerable people who set out to envision a future where the stars are far off marks of light as well as objections ready to be investigated. The tale of rockets trying the impossible is an account of human inventiveness, persistence, and the constant quest for the unexplored world.

In the next few decades, as new sections unfurl in the adventure of room investigation, the account will be formed by the choices we make today. The stars, when far off, entice us to wander forward with modesty and interest. The difficulties might be overwhelming, the excursion risky, however the prizes are unlimited. Rockets will keep on trying the impossible, leading of investigation that has enlightened the human soul all through the ages.

As we stand on the slope of another time in space investigation, let us recall the expressions of Carl Sagan: "The universe is inside us. We are made of star-stuff. We are a way for the universe to know itself." In the fantastic grandiose dance, let the rockets be our accomplices, impelling us toward a future where the stars are far off pinpricks in the velvet territory as well as objections that light the flares of creative mind and rouse ages yet unborn. The excursion has recently started, and the stars anticipate our appearance.

3.1 The Space Race: A competition between superpowers

The Space Race, a characterizing part throughout the entire existence of human investigation, unfurled against the scenery of Cold Conflict strains between two superpowers — the US and the Soviet Association. Starting in the last part of the 1950s, this extreme competition prodded a progression of historic accomplishments in space investigation, changing the fantasies about venturing the universe into a reality energized by political, mechanical, and philosophical inspirations.

The seeds of the Space Race were planted in the consequence of The Second Great War. The international scene moved as previous partners, the US and the Soviet Association, ended up on rival sides of the philosophical separation. The philosophical battle among a vote based system and socialism turned into a focal subject of the Virus War, penetrating each feature of global relations.

On October 4, 1957, the Soviet Association sent off the main counterfeit satellite, Sputnik 1, into space around the Earth. The blare blare of its radio transmissions stamped a logical victory as well as an emblematic second in the strengthening Cold Conflict contention. The effective send off of Sputnik exhibited the Soviet Association's innovative ability and surprised the US, making way for a space contest that would spellbind the world.

The shockwaves from Sputnik resounded through American culture and government. The apparent mechanical predominance of the Soviet Association started worries about public safety and America's remaining on the planet. Accordingly, the US escalated its endeavors to get up and outperform its Bug War foe in the domain of room investigation.

The space tries of the two superpowers turned out to be complicatedly connected to their international plans. The space race was not just a journey for logical information and mechanical progression; it was a fight for philosophical predominance. Each fruitful space mission turned into an emblematic triumph, exhibiting the prevalence of one framework over the other.

The Soviet Association proceeded with its dash of room achievements with the send off of the main living being, Laika, on board Sputnik 2 in November 1957. While this accomplishment brought up moral issues about the treatment of creatures in space, it hardened the Soviet Association's situation as the leader in space investigation.

Resulting Soviet missions accomplished extraordinary accomplishments, remembering the main human for space — Yuri Gagarin, who circled the Earth in April 1961 — and the primary lady in space — Valentina Tereshkova — in June 1963.

The US, under the initiative of President John F. Kennedy, answered the Soviet Association's space triumphs with a recharged obligation to space investigation. In a memorable discourse to Congress on May 25, 1961, Kennedy put forth an aggressive objective: to land an American on the moon and return them securely to Earth before the decade's end. This statement denoted the start of the Apollo program, a great endeavor that planned to defeat the difficulties of room travel and lunar investigation.

Apollo 11, the zenith of the Apollo program, accomplished its memorable mission on July 20, 1969. Space travelers Neil Armstrong and Buzz Aldrin turned into the principal people to go to the lunar surface, while Michael Collins circled above in the order module. The effective moon landing satisfied Kennedy's vision and exhibited American innovative ability, yet it likewise addressed a snapshot of worldwide importance. The expression "That is one little step for [a] man, one monster jump for humanity" reverberated all over the planet, rising above public limits and underlining the common human accomplishment in arriving at another heavenly body.

The US and the Soviet Association kept on contending in space investigation all through the 1960s and 1970s. The Apollo program saw a sum of six monitored moon arrivals, each adding to logical information and mechanical development. In the interim, the Soviet Association accomplished critical achievements with its Luna and Venera missions, which effectively investigated the moon and Venus, separately.

The Apollo-Soyuz Test Venture in 1975 denoted a representative defrost in Chilly Conflict pressures. American space travelers and Soviet cosmonauts docked their rocket in circle, exhibiting worldwide participation in space. While international competitions endured on The planet, the cooperative soul in space alluded to the potential for rising above natural struggles chasing shared objectives.

As the 1980s unfolded, the focal point of room investigation moved. The Space Transport program, started by NASA in 1981 with the send off of the Space Transport Columbia, planned to make space travel more everyday practice and financially savvy. The reusable rocket carried space explorers and payloads to and from low Earth circle, giving a stage to logical examination and satellite sending.

The finish of the Virus Battle in the last part of the 1980s and the resulting disintegration of the Soviet Association denoted a critical defining moment in the space race. The international scene went through a change, and the US arose as the sole superpower. With the disappearing of the philosophical competition that had energized the space race, questions emerged about the future heading of room investigation and the job of worldwide coordinated effort.

The post-Cold Conflict time saw a development in space collaboration. The Worldwide Space Station (ISS), a joint venture including the US, Russia, Europe, Japan, and Canada, turned into an image of global coordinated effort in space. Development of the ISS started in 1998, and it has filled in as a microgravity lab, encouraging logical exploration and participation among countries with different political foundations.

While the international inspirations of the Virus War time died down, new players entered the field of room investigation. The development of China as a spacefaring country in the 21st century added another dynamic to the worldwide space race. China's aggressive space program incorporated the send off of manned space missions, lunar investigation with the Chang'e missions, and the advancement of its own space station, Tiangong.

The job of privately owned businesses in space investigation acquired unmistakable quality, testing the conventional strength of government space organizations. Organizations like SpaceX, established by Elon Musk in 2002, altered space access with the improvement of the Bird of prey and Starship rocket frameworks. SpaceX's accomplishments, including the primary secretly evolved shuttle to moor with the ISS and the fruitful send off and return of reusable rocket stages, flagged another period in space business venture.

The space race of the 21st century reaches out past Earth's circle. Mars has turned into a point of convergence of investigation, with plans for maintained missions and the possible objective of laying out a human presence on the Red Planet. NASA's Tirelessness wanderer, sent off in 2020, keeps on investigating the Martian surface, looking for indications of previous existence and setting up the basis for future human missions.

As countries and confidential substances contend and team up in the cutting edge space race, inquiries concerning the maintainability of room exercises, the moral contemplations of planetary investigation, and the capable utilization of divine assets come to the front. The expanded presence of satellites in circle, worries about space flotsam and jetsam, and the potential for business abuse of divine bodies raise complex difficulties that require global participation and smart administration.

The space race, when characterized by the extreme contest between superpowers, has developed into a more different and cooperative undertaking. The inspirations for space investigation currently envelop logical disclosure, mechanical advancement, monetary open doors, and the intrinsic human drive to investigate the unexplored world. As countries and confidential substances put their focus on the moon, Mars, and then some, the tradition of the space race keeps on forming the fate of human space investigation.

In considering the historical backdrop of the space race, it is fundamental to perceive the significant effect it has had on society, innovation, and our comprehension of the universe. The rush to the moon catalyzed progressions in science and designing, moving ages of researchers, specialists, and visionaries. The innovative side projects from the Apollo program, going from worked on clinical gadgets to improved registering innovation, penetrated regular daily existence and added to the more extensive advancement of society.

In addition, the space race had international ramifications that reached out past the limits of the Earth. The opposition between the US and the Soviet Association in space was a sign of the more extensive philosophical battle of the Virus War. The victories and misfortunes in space were estimated in logical accomplishments as well as regarding political distinction, public pride, and worldwide impact.

In the ongoing period, space investigation has become more comprehensive, with a large number of countries and confidential substances effectively partaking. Worldwide coordinated efforts, exemplified by the ISS and joint missions to investigate the universe, highlight the potential for space to act as a bringing together power. While international pressures continue, the common undertaking of investigating the universe offers a road for participation that rises above natural divisions.

Looking forward, the space race keeps on unfurling on numerous fronts. Lunar investigation, with plans for ran missions and the foundation of lunar bases, has recaptured unmistakable quality. The Artemis program, drove by NASA, plans to return people to the moon and prepare for feasible lunar investigation. All the while, privately owned businesses imagine lunar the travel industry and asset usage, further broadening the partners in the new space race.

Mars, with its true capacity as an objective for human investigation and possible colonization, stays an enticing objective. The desires to arrive at the Red Planet include defeating huge difficulties, including long-term space travel, life emotionally supportive networks, and the advancement of economical territories. The fantasies about remaining on Martian soil, when the stuff of sci-fi, presently drive aggressive plans and missions.

The space race is not generally bound to government organizations; privately owned businesses are vital players in forming the fate of room investigation. SpaceX's aggressive objectives incorporate laying out a human presence on Mars, reforming space travel with reusable rockets, and conveying satellite star groupings for worldwide web inclusion. The combination of enterprising soul with space investigation can possibly change the financial aspects of room and open up new outskirts.

As humankind adventures further into the universe, the illustrations of the space race stay pertinent. The quest for aggressive objectives requires vision, assurance, and coordinated effort. The difficulties of room investigation request development, versatility, and a promise to pushing the limits of what is conceivable. The space race, with its celebrated history, fills in as a demonstration of the unstoppable human soul that tries to investigate, find, and try the impossible.

In the terrific account of the space race, from the beginning of Sputnik to the contemporary time of lunar and Martian aspirations, one ongoing idea ties the sections together — the unending mission for information, the persistent quest for the obscure, and the aggregate undertaking of mankind to investigate the universe. The space race isn't simply a verifiable relic however a living, developing story that keeps on molding the fate of our species in the boundlessness of the universe.

3.2 Yuri Gagarin's historic journey and the Apollo moon landings

The mid-twentieth century saw exceptional headways in space investigation, coming full circle in two noteworthy achievements that eternity modified mankind's impression of the universe — the main human in space, Yuri Gagarin, and the Apollo moon arrivals. These occasions, driven by political, logical, and philosophical inspirations, stamped crucial minutes in the space race between the US and the Soviet Association.

On April 12, 1961, Yuri Gagarin, a Soviet cosmonaut, set out on an excursion that would carve his name into the records of history. Sent off on board the space apparatus Vostok 1, Gagarin turned into the principal human to circle the Earth. The meaning of this accomplishment resounded internationally, rising above public limits and catching the minds of individuals all over the planet.

Gagarin's excursion into space was in excess of a mechanical accomplishment; it was an image of Soviet predominance in the expanding space race. The fruitful send off and safe return of the main human space traveler addressed a significant victory for the Soviet space program, which had proactively accomplished achievements with the send off of the principal counterfeit satellite, Sputnik, and the primary living being, Laika, into space.

The Vostok 1 mission endured only 108 minutes, during which Gagarin finished one circle around the Earth. His space apparatus arrived at a most extreme elevation of roughly 203 miles (327 kilometers). As Gagarin looked at the Earth from the bounds of his rocket, he radioed back the now-notable words, "Poyekhali!" signifying "We should go!" or "Off we go!" The mission finished up with Gagarin securely dropping back to Earth, arriving in the Saratov area of the Soviet Association.

The effect of Yuri Gagarin's process rose above the domain of room investigation. It was an international and philosophical victory for the Soviet Association during the level of the Virus War. The effective send off of the initial human into space exhibited the Soviet Association's mechanical ability and represented an immediate test to the US, which had its own yearnings for space investigation.

The US, under the administration of President John F. Kennedy, answered with a restored obligation to space investigation. Kennedy's renowned test, gave on May 25, 1961, during a discourse before a meeting of the US Congress, set up for a fantastic undertaking working closely together. He proclaimed, "I accept that this country ought to concede to accomplishing the objective, before this decade is out, of handling a man on the Moon and returning him securely to the Earth."

This aggressive objective turned into the main impetus behind the Apollo program, NASA's drive to land people on the Moon. The program confronted various difficulties, both specialized and calculated, as it tried to conquer the intricacies of room travel, lunar arrivals, and safe re-visitations of Earth. The rush to the Moon had started, with the US intending to accomplish an accomplishment that appeared to be nervy and nearly far-off.

On July 20, 1969, the zenith of long stretches of exertion, advancement, and penance happened when the Apollo 11 mission effectively landed two space travelers, Neil Armstrong and Buzz Aldrin, on the lunar surface. Armstrong's notable words as he slid the stepping stool of the lunar module, "That is one little step for [a] man, one goliath jump for humanity," reverberated across the globe. The accomplishment of putting people on the Moon and bringing them back securely addressed a victory of human inventiveness and resolve.

The Apollo 11 mission was a careful coordination of intricate moves. The rocket, comprising of the order module guided by Michael

Collins and the lunar module conveying Armstrong and Aldrin, had traveled more than 238,855 miles (384,400 kilometers) to arrive at the Moon. The lunar module, named Bird, isolated from the order module, Columbia, and plunged to the lunar surface.

The difficulties of the plummet were highlighted by the popular 1202 and 1201 cautions set off by the locally available PC. With the direction of mission control and fast reasoning by the space explorers and specialists, the lunar module effectively arrived in the Ocean of Serenity, a locale on the Moon's surface.

Armstrong and Aldrin spent roughly over two hours directing investigations, gathering tests, and taking photos on the lunar surface. The noteworthy moonwalk was seen by millions all over the planet, denoting the acknowledgment of President Kennedy's vision and the satisfaction of a guarantee to mankind.

While Armstrong and Aldrin investigated the lunar surface, Michael Collins circled above in the order module, anticipating their return. The rising phase of the lunar module rendezvoused with the order module, and the space travelers moved back to Columbia for the excursion back to Earth. The fruitful splashdown in the Pacific Sea on July 24, 1969, denoted the victorious finish of the Apollo 11 mission.

The Apollo moon arrivals went on with resulting missions, each structure upon the information and experience acquired from its ancestors. Apollo 12, 14, 15, 16, and 17 broadened the human presence on the lunar surface, leading analyses, sending logical instruments, and extending how we might interpret Earth's divine buddy.

The logical tradition of the Apollo program is significant. The lunar missions brought back an abundance of geographical examples, offering experiences into the Moon's organization and history. The tests led on the lunar surface gave important information, and the lunar seismometers set during Apollo missions distinguished moonquakes, revealing insight into the Moon's inside structure.

In any case, the Apollo program confronted analysis for its expense and the discernment that it was driven more by political contemplations than logical investigation. As the 1970s unfurled, monetary requirements and changing needs prompted the undoing of arranged Apollo missions, and human lunar investigation was briefly required to be postponed.

The tradition of Yuri Gagarin and the Apollo moon arrivals reaches out past the logical and mechanical accomplishments. These occasions hold social and emblematic importance that rises above their nearby setting. Gagarin's excursion into space turned into an image of Soviet greatness, showing the way that people could wander past Earth's climate and get back securely.

In the US, the Apollo program represented the country's mechanical ability and its ability to defeat apparently impossible difficulties. The picture of the American banner established on the lunar surface turned into a persevering through image of human accomplishment and the unstoppable soul of investigation.

The space race between the US and the Soviet Association, portrayed by the noteworthy excursions of Yuri Gagarin and the Apollo space explorers, had significant international ramifications. The opposition stretched out past the limits of Earth, venturing into the limitlessness of room. The accomplishments in space became markers of public eminence and mechanical authority during when the Virus War contention was at its pinnacle.

The space race was a mechanical challenge as well as a clash of belief systems. The US, upholding for a majority rule government and unregulated economy standards, and the Soviet Association, supporting socialism, looked to exhibit the predominance of their individual frameworks. The moon arrivals, specifically, turned into a feature for the popularity based goals of investigation, opportunity, and human resourcefulness.

While the Virus War contention gave the underlying driving force to the space race, the tradition of Yuri Gagarin and the Apollo moon arrivals goes past international contemplations.

These accomplishments motivated ages of researchers, architects, and visionaries all over the planet. The possibility that people could wander past their home planet and investigate other heavenly bodies turned into a strong image of human potential and the constant quest for information.

The social effect of these memorable occasions is apparent in human expression, writing, and mainstream society. The notorious symbolism of space travelers strolling on the moon and the Earthrise seen from the lunar surface became imbued in the shared mindset. Movies, books, and fine arts have drawn motivation from the glory and meaning of human space investigation.

In the many years that followed, space investigation developed. The space transport time, which started during the 1980s, saw reusable rocket giving routine admittance to low Earth circle. The development and activity of the Global Space Station (ISS), a cooperative exertion including numerous countries, denoted another part in human space investigation. The ISS filled in as a microgravity lab, cultivating logical examination and global participation.

As the 21st century unfurled, new players entered the field of room investigation. Privately owned businesses, like SpaceX and Blue Beginning, became fundamental to the space business, testing customary models and driving development. The center moved toward lunar investigation, with plans for maintainable lunar environments, ran missions, and the Artemis program intending to return people to the Moon.

Yuri Gagarin and the Apollo space travelers, as pioneers in space investigation, established the groundwork for ensuing undertakings. Their fortitude, versatility, and obligation to pushing the limits of human capacity propelled the up and coming age of room travelers. The soul of investigation, epitomized in these notable excursions, stays a main impetus as humankind looks toward future endeavors into profound space.

The tradition of Yuri Gagarin and the Apollo moon arrivals isn't bound to the past. It lives on in the goals of current and future space missions. The vision of sending people to Mars, the investigation of space rocks, and the mission to find livable exoplanets are all essential for a continuous account that expands upon the accomplishments of the space race.

3.3 Technological advancements in rocket design and engineering during the mid-20th century

The mid-twentieth century denoted a groundbreaking period in the field of rocket plan and designing, driven by remarkable logical and mechanical headways. This period, molded by the consequence of The Second Great War and the beginning of the Virus War, saw the quick improvement of rocket innovation that would at last move humankind into the space age. The central members in this mechanical race were the US and the Soviet Association, each competing for matchless quality in space investigation.

The foundations of mid-twentieth century rocketry can be followed back to the spearheading work of visionaries like Konstantin Tsiolkovsky and Robert H. Goddard. Tsiolkovsky, a Russian researcher, laid the hypothetical foundation for rocket impetus in the mid twentieth 100 years. His visionary thoughts, including the idea of utilizing multistage rockets to arrive at orbital speeds, roused people in the future of technical geniuses. In the mean time, Goddard, an American physicist, led useful examinations during the 1920s and 1930s, effectively sending off the world's most memorable fluid filled rocket in 1926.

The finish of The Second Great War saw the rise of caught German rocket innovation as a sought after prize for the Partnered powers. German researchers, drove by figures like Wernher von Braun, had fostered the V-2 rocket — a progressive weapon with the capacity to arrive at the edge of room prior to sliding to its objective. The V-2 addressed a jump forward in rocket designing, highlighting developments like fluid energized motors, gyroscopic direction, and smoothed out streamlined features.

The mechanical race for rocket matchless quality started vigorously with the beginning of the Virus War. The international pressures between the US and the Soviet Association gave the driving force to quick headways in rocket plan and designing. The Soviets, utilizing the mastery of German researchers brought to the Soviet Association under Activity Paperclip, immediately retained and developed V-2 innovation.

In 1957, the Soviet Association accomplished a memorable achievement that resonated across the globe — the send off of Sputnik 1, the main fake satellite. This occasion checked a victory in space investigation as well as a show of the Soviet Association's dominance of rocket innovation. Sputnik 1, controlled by a changed R-7 intercontinental long range rocket (ICBM), circled the Earth and discharged radio transmissions that proclaimed the beginning of the space age.

The outcome of Sputnik 1 was immediately trailed by another huge accomplishment — the send off of the initial living being into space. Laika, a Soviet space canine, circled the Earth on board Sputnik 2 in November 1957. While Laika's central goal brought up moral issues about the treatment of creatures in space, it highlighted the Soviet Association's obligation to pushing the limits of room investigation.

In light of the Soviet triumphs, the US heightened its endeavors to get up and outperform its Bug War enemy. The foundation of the Public Flying and Space Organization (NASA) in 1958 denoted a urgent second in American space investigation. NASA's order included logical investigation as well as public safety contemplations, especially because of the Soviet Association's accomplishments in rocketry.

The Redstone and Chart book rockets became fundamental parts of the US's initial space tries. The Redstone rocket, an immediate relative of the V-2, was utilized to send off the initial American into space — Alan Shepard — during the Mercury-Redstone 3 mission in 1961. The Map book rocket, initially created as an ICBM, was adjusted for space investigation and turned into the send off vehicle for the Mercury and Gemini programs.

The Mercury and Gemini programs, NASA's initial manned space missions, assumed a vital part in testing the capacities of the two space travelers and the rockets that would impel them into space. These projects gave important bits of knowledge into the difficulties of human spaceflight, including orbital mechanics, life emotionally supportive networks, and the physiological impacts of broadened space travel.

As the 1960s advanced, the US put its focus on the aggressive objective set by President John F. Kennedy — handling a man on the Moon and returning him securely to Earth before the decade's end. The Apollo program, NASA's lunar investigation drive, required the improvement of strong new send off vehicles fit for conveying space travelers and their space apparatus past Earth's air and on a direction toward the Moon.

The Saturn group of rockets turned into the workhorse of the Apollo program. The Saturn I and IB rockets were utilized for early ran space missions, while the Saturn V, one of the most remarkable rockets at any point fabricated, would convey space travelers to the Moon. The improvement of the Saturn rockets requested developments in drive, materials, and designing to accomplish the essential push and payload limit.

The Saturn V, remaining at north of 363 feet (110 meters) tall, included three phases, each controlled by strong rocket motors. The primary stage utilized five F-1 motors, creating a consolidated push of over 7.5 million pounds. The subsequent stage utilized five J-2 motors, while the third stage used a solitary J-2 motor to move the space apparatus into lunar circle.

The Apollo 11 mission, the primary effective lunar landing, depended on the Saturn V rocket to move space explorers Neil Armstrong, Buzz Aldrin, and Michael Collins to the Moon. The titanic rocket performed perfectly, sending off the space apparatus into space and later infusing it onto a direction toward the Moon. The accuracy and unwavering quality of the Saturn V were basic elements in the progress of the Apollo 11 mission.

The improvement of the Saturn rockets and the ensuing lunar arrivals addressed a peak in mid-twentieth century rocket plan and designing. The difficulties of defeating Earth's gravity, accomplishing the important speeds for lunar missions, and guaranteeing the protected return of space travelers requested unrivaled accuracy and development. The examples gained from the Apollo program established the groundwork for future space investigation tries.

In lined up with the maintained space missions, mechanical space tests assumed a pivotal part in propelling rocket innovation and extending how we might interpret the nearby planet group. Tests like the Trailblazer and Explorer series, sent off during the 1960s and 1970s, depended on strong rockets to get away from Earth's gravitational draw and set out on interplanetary excursions. These missions directed flybys, accumulated important information about the external planets, and eventually wandered into the most distant scopes of our planetary group.

Mechanical progressions during the mid-twentieth century were not restricted to drive frameworks; materials science assumed a urgent part in rocket plan. The requirement for lightweight yet solid materials prompted the advancement of new compounds and composites fit for enduring the extraordinary anxieties of room travel. The utilization of lightweight materials further developed eco-friendliness and payload limit, adding to the general proficiency of rocket frameworks.

The mid-twentieth century likewise saw progressions in direction and control frameworks fundamental for precise and solid space missions. Inertial route frameworks, gyrators, and modernized control frameworks empowered exact direction changes and guaranteed that rockets could arrive at their planned objections with astounding precision.

The space transport time, which started in the mid 1980s, presented new advancements in rocket plan. The Space Transport, with its reusable orbiter and strong rocket promoters, addressed a takeoff from conventional disposable send off vehicles. The bus' primary motors, powered by fluid hydrogen and fluid oxygen, gave a serious level of productivity and reusability.

Regardless of the mechanical accomplishments of the space transport, worries about wellbeing and cost at last prompted its retirement in 2011. The resulting movement to new send off vehicles, including privately owned businesses like SpaceX presenting reusable rocket innovation, denoted another stage in space investigation. The Hawk and Starship rockets created by SpaceX meant to additionally lessen the expenses related with space travel and increment the recurrence of dispatches.

The appearance of business space investigation, drove by organizations like SpaceX, Blue Beginning, and others, has carried recharged regard for rocket plan and designing. The drive for financially savvy and manageable admittance to space has prodded developments in impetus, materials, and assembling processes. Reusable rocket stages, vertical landing capacities, and progressions in independent direction frameworks are among the vital improvements in contemporary rocket plan.

Drive innovations have developed too, with an emphasis on expanding proficiency and diminishing ecological effect. Electric drive, especially particle and Lobby impact engines, is turning out to be more predominant for profound space missions, offering higher explicit motivation and eco-friendliness contrasted with conventional substance impetus frameworks.

All in all, the mid-twentieth century saw an upheaval in rocket plan and designing that pushed mankind into the space age. The international rivalry between the US and the Soviet Association, joined with the visionary work of early trailblazers like Tsiolkovsky and Goddard, established the groundwork for uncommon progressions in drive, materials science, and direction frameworks.

The V-2 rocket's heritage, caught German innovation, and the ensuing improvement of strong send off vehicles like the Saturn series prepared for notable accomplishments, for example, the Apollo moon arrivals. The space race displayed human inventiveness as well as sped up the speed of mechanical development, making a permanent imprint on the direction of room investigation.

Chapter 4

Unleashing the Power: Rocket Propulsion Technologies

Rocket drive advancements play had a vital impact in molding the course of human investigation and mechanical headway. From the beginning of simple rocketry to the refined drive frameworks controlling present day space missions, the excursion has been completely remarkable. The steady quest for arriving at the stars has driven researchers, architects, and visionaries to push the limits of what is conceivable, prompting notable advancements in rocket impetus.

One of the earliest occurrences of rocketry traces all the way back to old China, where explosive filled tubes were utilized for military purposes. These early trials established the groundwork for future advancements in rocket innovation. In any case, it was only after the twentieth century that rockets started to become the dominant focal point in the domain of room investigation.

The spearheading work of visionaries like Konstantin Tsiolkovsky, Robert H. Goddard, and Hermann Oberth prepared for the improvement of fluid filled rockets. Tsiolkovsky, a Russian researcher, is frequently viewed as the "father of astronautics" for his notable hypothetical work on rocket drive. Goddard, an American physicist, is credited with building the world's most memorable fluid energized rocket in 1926. Oberth, a German researcher, made huge commitments to the hypothetical comprehension of rocketry.

WWII denoted a defining moment for rocket innovation, with both the Partners and Hub powers putting vigorously in the improvement of long range rockets. The V-2 rocket, created by Nazi Germany, turned into the world's most memorable long-range directed long range rocket. The finish of the conflict saw a flood in logical coordinated effort and the deluge of German technical geniuses into the US and the Soviet Association, making way for the space race.

The Virus War contention between the US and the Soviet Association filled fast headways in rocket drive. The send off of the Soviet satellite Sputnik 1 of every 1957 denoted the start of the space age, setting off a competition to investigate space. The US answered with the fruitful send off of Pilgrim 1 of every 1958. This period saw the improvement of an assortment of rocket drive innovations, including the utilization of fluid, strong, and half breed fuels.

Fluid rocket motors turned into the workhorses of room investigation during the beginning of the space race. These motors depend on fluid charges, ordinarily a fluid fuel and an oxidizer, which are put away in isolated tanks and blended in a burning chamber to create push. The Saturn V rocket, utilized in the Apollo moon missions, highlighted strong fluid rocket motors that moved space explorers past Earth's air.

Strong rocket drive, then again, includes the utilization of a strong combination of fuel and oxidizer. Strong rocket engines are known for their straightforwardness and dependability, making them reasonable for different applications, including military rockets and space send off vehicles. The Space Transport's strong rocket sponsors, for instance, gave extra push during the underlying period of send off.

Half breed rocket impetus joins components of both fluid and strong drive. In mixture frameworks, a fluid oxidizer is ordinarily utilized related to a strong fuel. This arrangement offers benefits, for example, exact control of pushed and the capacity to close down and restart the motor. Virgin Cosmic's SpaceShipOne, the primary secretly evolved monitored shuttle, used a half and half rocket engine for sub-orbital spaceflight.

Progressions in materials science, producing procedures, and computational demonstrating have additionally upgraded the exhibition and productivity of rocket drive frameworks. High-strength amalgams, lightweight composites, and high level ceramics are utilized to endure the outrageous circumstances experienced during send off and space travel.

Added substance fabricating, regularly known as 3D printing, has altered the creation of intricate motor parts, diminishing expenses and lead times.

The journey for higher effectiveness and expanded payload limit has prompted the improvement of creative drive ideas. Particle drive, for instance, uses electric or electromagnetic fields to speed up particles, delivering exhaust speeds a lot higher than conventional compound rockets. While particle motors produce low push, their proficiency over lengthy term missions makes them ideal for specific space investigation missions.

Atomic warm drive addresses one more boondocks in rocket innovation. This idea includes utilizing an atomic reactor to warm a charge, regularly hydrogen, to high temperatures, bringing about a high velocity exhaust that produces push. The potential for essentially higher explicit motivation contrasted with synthetic rockets makes atomic warm drive a convincing choice for maintained missions to Mars and then some.

Lately, privately owned businesses, for example, SpaceX play had an extraordinary impact in the space business, driving development and diminishing send off costs. SpaceX's Bird of prey 9 and Hawk Weighty rockets, fueled by the Merlin motors, epitomize the mix of state of the art innovation and business suitability. Reusable rocket parts, for example, the main stage promoters that can be recuperated and relaunched, have turned into a distinct advantage in the financial matters of room travel.

The rise of new players in the space business has ignited a renaissance in rocket drive innovative work. Blue Beginning, established by Amazon's Jeff Bezos, is effectively chasing after the advancement of reusable rocket motors and the foundation of business space the travel industry. The opposition and joint effort between organizations like SpaceX and Blue Beginning are driving quick progressions in impetus advancements, pushing the limits of what was once considered unthinkable.

In the domain of room investigation, impetus isn't just about arriving at objections yet additionally about guaranteeing the security and prosperity of space travelers during their excursions. Life emotionally supportive networks, radiation protecting, and high level instrumentation are necessary parts of maintained rocket. Drive frameworks should be solid, productive, and fit for supporting long-length missions to objections like Mars, where travel times can stretch out for quite a long time.

As mankind looks past Earth's circle, the improvement of cutting edge impetus advances becomes vital. Ideas, for example, the EmDrive, a dubious impetus framework that purportedly produces push without the requirement for fuel, have created both energy and incredulity inside mainstream researchers. While trial results have shown minute pushed levels, the basic physical science of the EmDrive stay a subject of discussion.

Hypothetical ideas, for example, the Alcubierre drive propose a twist bubble in spacetime, considering quicker than-light travel. Nonetheless, the difficulties related with controlling spacetime and the requirement for extraordinary types of issue raise huge obstacles for the functional acknowledgment of such ideas. Notwithstanding the speculative idea of these thoughts, they feature the inventive and ground breaking nature of impetus research.

The ecological effect of rocket dispatches has likewise gone under investigation as space exercises increment. Customary rocket charges discharge huge amounts of carbon dioxide and different poisons into the air. As the recurrence of dispatches rises, tending to the ecological impression of room travel becomes basic. Specialists are investigating greener force choices and more reasonable send off practices to moderate the natural effect of room investigation.

The fate of rocket impetus holds invigorating potential outcomes, from the possibility of interstellar travel to the colonization of other heavenly bodies. Forward leaps in materials science, impetus ideas, and space investigation strategies are driving the advancement of rocketry. The Artemis program, drove by NASA, means to return people to the Moon and prepare for future ran missions to Mars. The improvement of the Space Send off Framework (SLS) and the Orion shuttle addresses another section in human space investigation.

Global cooperation is progressively turning into a sign of room investigation endeavors. Organizations between space offices, privately owned businesses, and worldwide associations are cultivating a common vision for the eventual fate of room investigation. The Global Space Station (ISS) remains as a demonstration of the cooperative endeavors of different countries in propelling human spaceflight capacities and leading logical examination in microgravity.

As we ponder the eventual fate of rocket impetus innovations, contemplations of moral, lawful, and cultural effects come to the front. The mindful utilization of room assets, the counteraction of room trash, and the evenhanded circulation of advantages from space investigation are necessary parts of a maintainable spacefaring future. The Space Settlement, laid out in 1967, gives a system to overseeing the exercises of countries in space and highlights the significance of global collaboration.

All in all, the excursion of rocket drive innovations from old Chinese black powder filled cylinders to current space investigation addresses a wonderful odyssey of human creativity and assurance. The union of logical standards, designing ability, and visionary desire has pushed humankind past the limits of Earth, opening new wildernesses for investigation and revelation.

The continuous progressions in rocket impetus, driven by both government space organizations and confidential elements, guarantee a future where space travel is more open, productive, and supportable. From reusable rocket parts to inventive impetus ideas, the scene of room investigation is advancing at an uncommon speed.

4.1 In-depth exploration of different propulsion systems

The investigation of room has been a longstanding human undertaking, and impetus frameworks have been at the front of empowering this journey. As we dig into an inside and out investigation of various impetus frameworks, we experience a different cluster of innovations that have developed over the long haul, each with its novel benefits and difficulties.

Conventional substance rocket impetus has been the workhorse of room investigation for quite a long time. This strategy depends on the ignition of synthetic fuels to create push. The straightforwardness and dependability of compound rockets make them appropriate for a great many applications, from sending off satellites into space to sending manned missions to the Moon and then some.

Fluid rocket motors are a vital subset of synthetic impetus, utilizing fluid powers and oxidizers that are put away in isolated tanks and blended in a burning chamber. The controlled start of these fluids delivers a high velocity exhaust that moves the rocket forward. The Apollo missions, including the notable Apollo 11 moon landing, used fluid rocket motors to accomplish the important push for getting away from Earth's gravitational draw.

Strong rocket drive addresses one more part of synthetic impetus, where a strong combination of fuel and oxidizer is preloaded into the rocket engine. Upon start, the strong charge goes through burning, delivering push. Strong rocket engines are known for their straightforwardness and unwavering quality, making them reasonable for military applications, supporter stages, and, surprisingly, the Space Transport's strong rocket promoters.

Mixture rocket impetus consolidates components of both fluid and strong drive. In a normal cross breed framework, a fluid oxidizer is matched with a strong fuel. This arrangement offers benefits, for example, exact control of pushed and the capacity to close down and restart the motor. Virgin Cosmic's SpaceShipOne, which finished the principal secretly created monitored spaceflight, used a half and half rocket engine.

Progressions in materials science and designing play had a vital impact in improving the presentation and effectiveness of customary synthetic rockets. High-strength compounds and lightweight composites are utilized to endure the outrageous states of room travel. Besides, the improvement of cutting edge fabricating strategies, including added substance fabricating, has smoothed out the creation of intricate rocket parts.

While substance rockets have been the essential method for arriving at space, their limits become clear while thinking about the difficulties of profound space investigation. The requirement for a lot of charge and the generally low unambiguous drive, a proportion of impetus proficiency, make compound rockets less reasonable for long-term missions past Earth's circle.

Because of these difficulties, analysts and specialists have investigated elective drive innovations to push the limits of room investigation. Particle impetus, otherwise called electric drive, addresses one such development. Particle motors utilize electric or electromagnetic fields to speed up particles (typically xenon) to create push. Despite the fact that particle motors produce low push, they accomplish a lot higher exhaust speeds contrasted with synthetic rockets, bringing about more noteworthy eco-friendliness overstretched mission spans.

NASA's Day break space apparatus, which investigated the bantam planets Ceres and Vesta, highlighted particle drive. The supported push given by particle motors permitted First light to embrace a mind boggling direction, visiting various divine bodies in the space rock belt.

While particle impetus is great for long-term missions, its low pushed limits its appropriateness for quick moves and dispatches from Earth.

Atomic warm drive is an idea that holds guarantee for future maintained missions to Mars and then some. This impetus framework use an atomic reactor to warm a force, regularly hydrogen, to high temperatures. The rapid exhaust delivered by the warmed charge produces push. Atomic warm impetus offers fundamentally higher explicit motivation contrasted with synthetic rockets, empowering quicker travel times and more proficient utilization of charge.

In spite of the likely benefits of atomic warm impetus, difficulties, for example, the advancement of hearty reactor frameworks, radiation protecting, and the protected treatment of atomic materials should be survived. Hypothetical ideas like the Atomic Warm Rocket (NTR) have been proposed for maintained missions to Mars, where diminishing travel time is basic for the prosperity of space explorers.

Another modern impetus idea that has gathered consideration is the EmDrive (Electromagnetic Drive). Not at all like conventional rocket motors that oust fuel to produce push, the EmDrive purportedly creates push without the requirement for charge ejection. The idea depends on electromagnetic reverberation inside a tightened depression to deliver a net push. Nonetheless, the EmDrive has been met with distrust inside established researchers, as it apparently disregards the protection of energy.

Tests testing the EmDrive have detailed minute degrees of pushed, yet the fundamental material science stay a subject of discussion. Pundits contend that the noticed push might be credited to exploratory mistakes or outer factors as opposed to a veritable drive impact. Notwithstanding the wariness, the EmDrive represents the quest for whimsical impetus ideas that challenge how we might interpret traditional physical science.

The Alcubierre drive addresses a hypothetical idea that charms the creative mind with its true capacity for quicker than-light travel. Proposed by physicist Miguel Alcubierre, the thought includes making a twist bubble in spacetime that agreements space before a rocket and extends it behind, successfully permitting the space apparatus to ride a flood of packed space.

While the Alcubierre drive is predictable with general relativity, the useful acknowledgment of such an idea faces huge obstacles, including the requirement for intriguing types of issue with negative energy thickness.

As of late, the business space industry has seen a change in outlook with the development of privately owned businesses playing a main job in space investigation. SpaceX, established by Elon Musk, has been at the very front of this change. The Bird of prey 9 and Hawk Weighty rockets, controlled by the Merlin motors, have shown the feasibility of reusable rocket innovation.

The reusability of rocket parts, especially the primary stage promoters, can possibly reform the financial matters of room travel. Recuperating and revamping these parts altogether lessens send off costs, making space access more reasonable. SpaceX's aggressive Starship project intends to additional push the limits of reusability and work with missions to the Moon, Mars, and then some.

Blue Beginning, established by Jeff Bezos, is one more central member in the confidential space area. Blue Beginning's New Shepard suborbital rocket, intended for space the travel industry, has finished different effective practice runs. The organization is likewise fostering the New Glenn orbital rocket and the BE-4 rocket motor, which is expected for use in different send off vehicles.

These confidential endeavors address a powerful power driving development in rocket impetus advancements. The cutthroat scene encourages fast headways, as organizations compete for agreements and organizations with government space offices. The interaction among public and confidential substances is forming the eventual fate of room investigation, with an accentuation on cost-adequacy, productivity, and supportability.

Headways in impetus advancements are naturally connected to the more extensive objectives of room investigation, including the mission for tenable conditions past Earth. Manned missions to the Moon and Mars require effective impetus frameworks as well as life emotionally supportive networks, radiation safeguarding, and practical territories. The Artemis program, drove by NASA, intends to return people to the Moon and lay out an economical human presence.

Worldwide cooperation is turning out to be progressively urgent in space investigation endeavors. The Worldwide Space Station (ISS) fills in as an image of worldwide collaboration, uniting space explorers, researchers, and specialists from different countries. The trading of information and assets on the ISS has added to progressions in space innovation and logical examination in a microgravity climate.

The Space Deal, laid out in 1967, gives a legitimate system to overseeing the exercises of countries in space. Key standards incorporate the quiet utilization of space, the counteraction of weaponization, and the aversion of hurtful defilement. As space exercises increment, the significance of peaceful accords and standards becomes central in guaranteeing the dependable and feasible utilization of room assets.

The ecological effect of rocket dispatches has likewise gone under investigation. Conventional rocket fuels discharge enormous amounts of carbon dioxide and different poisons into the air, adding to environmental change. As the recurrence of dispatches rises, scientists and space organizations are investigating greener force choices and more economical send off rehearses.

Green charges, for example, those in view of hydroxylammonium nitrate, offer decreased poisonousness and ecological effect contrasted with customary hypergolic fuels. The turn of events and reception of greener forces line up with the more extensive objectives of alleviating the natural impression of room investigation.

All in all, the investigation of various impetus frameworks uncovers a rich embroidery of mechanical development, from the tried and true dependability of synthetic rockets to the modern ideas testing the limits of material science. The development of impetus innovations is firmly entwined with the more extensive story of human space investigation, driven by a tireless mission for information, disclosure.

4.2 Liquid and solid rocket fuels

The domain of rocketry is pushed by the principal decisions of rocket energizes, with fluid and strong charges remaining as the two essential classes that have fueled space investigation and military applications for a really long time. Each sort of charge has its unmistakable qualities, benefits, and restrictions, forming the direction of rocket configuration, space investigation, and satellite send-offs.

Fluid Rocket Fills:

Fluid rocket fills are a foundation of room investigation, highlighting noticeably in the send off vehicles that have moved satellites, ran missions, and logical payloads into space. The critical qualification of fluid rocket fills lies in their organization — these charges comprise of fluid oxidizers and fluid powers that are put away in discrete tanks until the snapshot of start.

One of the early trailblazers of fluid rocketry was Robert H. Goddard, who effectively sent off the world's most memorable fluid energized rocket in 1926. This obvious a crucial second in the improvement of rocket impetus, moving from the straightforwardness of strong forces to the expanded control and proficiency managed the cost of by fluid fills.

Fluid rocket motors work on the rule of controlled ignition. The fluid fuel and oxidizer are siphoned into a burning chamber, where they light after blending. The subsequent high-pressure, high-temperature gases are removed through a rocket spout, creating push as per Newton's third law of movement.

An eminent illustration of fluid rocket innovation is the well known Saturn V rocket, which fueled the Apollo missions to the Moon. The F-1 motors on the Saturn V used fluid oxygen (LOX) as the oxidizer and lamp oil as the fuel. The productive mix of these charges gave the enormous push expected to get away from Earth's gravity and attempt the excursion to the Moon.

Hydrogen and oxygen have additionally been generally utilized as fluid rocket fuels. The Space Transport's principal motors, for example, utilized fluid hydrogen and fluid oxygen in an organized burning cycle. This mix brought about high unambiguous drive, a proportion of impetus productivity, setting it reasonable for the expectations of room travel.

Notwithstanding their adequacy, fluid rocket motors have specific disadvantages. The intricacy of the pipes and control frameworks expected for taking care of fluid charges expands the general intricacy of the rocket. Also, the requirement for protection and cooling components for cryogenic charges adds to the designing difficulties. In any case, the benefits as far as exact control, the capacity to close down and restart motors, and higher explicit drive have set the significance of fluid rocket energizes in space investigation.

Strong Rocket Powers:

As opposed to fluid rocket energizes, strong rocket forces are described by a solitary, homogeneous combination of fuel and oxidizer. This sythesis is preloaded into the rocket engine, framing a strong mass. When touched off, ignition proliferates through the whole charge grain, delivering push.

Strong rocket engines are known for their effortlessness and unwavering quality. They are normally utilized in military applications, like rockets and gunnery, as well as in promoter phases of send off vehicles. The Space Transport, for instance, used two strong rocket supporters to give extra push during takeoff.

The piece of strong rocket forces normally incorporates powdered metals, a powdered oxidizer, and a folio to keep the blend intact. Ammonium perchlorate is an ordinarily utilized oxidizer, while powdered aluminum fills in as the fuel. The fastener, frequently hydroxyl-ended polybutadiene (HTPB), gives underlying honesty to the charge grain.

The straightforwardness of strong rocket engines adds to their unwavering quality, making them appropriate for specific applications. Be that as it may, their powerlessness to be choked or closed down once lighted limits their adaptability.

When touched off, a strong rocket engine will consume until the whole fuel grain is exhausted, which can be a detriment in situations requiring exact control.

Notwithstanding these constraints, the expense adequacy and dependability of strong rocket engines have prompted their proceeded with use in different applications. Military applications, where quick reaction and effortlessness are pivotal, frequently favor strong rocket impetus. Also, the effortlessness of configuration makes strong rocket sponsors reasonable for giving introductory push during the send off of bigger vehicles.

Similar Investigation:

The decision among fluid and strong rocket energizes relies upon the particular prerequisites of the mission, the ideal degree of control, and the compromises between intricacy, cost, and dependability. Fluid rocket motors succeed in circumstances where exact control of pushed and the capacity to close down and restart motors are basic.

The adaptability of fluid rocket motors likewise considers the utilization of different fuels, each offering explicit benefits. Hypergolic fuels, which touch off precipitously upon contact, wipe out the requirement for a start framework. Notwithstanding, their harmfulness and taking care of difficulties have prompted the investigation of greener other options. The advancement of green fuels, for example, hydroxylammonium nitrate-based details, expects to address natural worries related with conventional hypergolic charges.

Then again, strong rocket engines find their specialty in applications where effortlessness, cost-adequacy, and unwavering quality are principal. Military rockets, space send off promoters, and certain strategic applications benefit from the fast reaction and clear plan of strong rocket drive. The failure to close down or choke strong rocket engines is a constraint that mission organizers should cautiously consider.

Progressions in materials science and designing have added to working on the exhibition of both fluid and strong rocket fuels. High-strength amalgams, composite materials, and high level assembling methods have upgraded the primary respectability of fluid rocket parts. For strong rocket engines, advancements in charge details and projecting advancements have prompted enhancements in execution and wellbeing.

Future Patterns and Advancements:

Looking forward, the field of rocket impetus is seeing continuous advancements and improvements that intend to address current restrictions and open additional opportunities. The journey for additional productive and harmless to the ecosystem fuels proceeds, with analysts investigating options that balance execution, security, and manageability.

Crossover rocket drive, which consolidates components of both fluid and strong impetus, is an area of dynamic exploration. Mixture frameworks ordinarily include a fluid oxidizer and a strong fuel, offering the benefits of exact control and the capacity to close down and restart the motor. This arrangement presents potential open doors for further developed wellbeing and execution in contrast with customary strong rocket engines.

Headways in added substance fabricating, usually known as 3D printing, are changing the scene of rocket part creation. The capacity to produce complex and streamlined structures utilizing 3D printing adds to decreased expenses and lead times. Both fluid and strong rocket parts stand to profit from the proceeded with advancement of added substance fabricating innovations.

In the journey for higher proficiency and execution, atomic warm impetus stays a convincing idea for future maintained missions past Earth. The utilization of atomic reactors to warm fuels, for example, hydrogen offers the potential for essentially higher explicit drive contrasted with compound rockets. In spite of specialized difficulties, the improvement of atomic warm impetus frameworks is a subject of restored interest.

In the domain of room investigation, the quest for fuel choices that limit the natural effect of rocket dispatches is acquiring noticeable quality. Green forces, portrayed by decreased poisonousness and lower ecological perils, are overall effectively investigated and tried. As the recurrence of room missions builds, the reception of harmless to the ecosystem forces lines up with more extensive endeavors to advance supportability in space investigation.

4.3 Innovations in propulsion efficiency and sustainability

In the unique scene of room investigation, drive productivity and maintainability have become central focuses for development. As humankind keeps on pushing the limits of our vast undertakings, headways in impetus advances are fundamental for making space travel more proficient, practical, and harmless to the ecosystem. This investigation digs into the developments that drive impetus proficiency and manageability, analyzing the advancements that guarantee to reshape the eventual fate of room investigation.

Electric Drive:

One of the vital developments in impetus productivity is electric drive, otherwise called particle impetus. Customary substance rockets create push by ousting force at high velocities, adhering to Newton's third regulation. Conversely, electric impetus depends on the speed increase of particles utilizing electric or electromagnetic fields. While electric motors produce lower push contrasted with synthetic rockets, they accomplish essentially higher exhaust speeds, bringing about higher explicit motivation and eco-friendliness over lengthy span missions.

The Corridor impact engine is a remarkable illustration of electric drive. It uses an attractive field to ionize a fuel, normally xenon, and ousts the particles to create push. Particle motors have been effectively utilized in different space missions, for example, NASA's Day break space apparatus, which investigated the space rock belt, and the European Space Office's (ESA) BepiColombo mission to Mercury.

The productivity acquires given by electric drive are especially profitable for profound space missions that require expanded times of pushed. This innovation is instrumental in decreasing how much fuel required for long excursions, settling on it a convincing decision for interplanetary investigation and logical missions to the external spans of the planetary group.

Reusability:

In the mission for supportability and cost-viability, the idea of reusability has arisen as a groundbreaking development in rocket plan. By and large, most rocket parts were viewed as nonessential, disposed of after a solitary use send off. In any case, spearheading endeavors by privately owned businesses, like SpaceX, have shown the plausibility and financial benefits of reusable rocket parts.

SpaceX's Bird of prey 9 and Hawk Weighty rockets have set the benchmark for reusability. The primary stage sponsors of these rockets are intended to get back to Earth after send off, landing upward for renovation and reuse in ensuing missions. This weighty methodology has prompted significant expense reserve funds, as the significant expense of a conventional rocket send off lies in the development of new rocket parts.

The outcome of reusable rocket innovation has reshaped the financial matters of room travel, making it more available and economically reasonable. Organizations like Blue Beginning, established by Jeff Bezos, have additionally embraced reusability in their rocket plans. The New Shepard suborbital rocket, expected for space the travel industry, includes a reusable supporter that goes through various flights.

The shift towards reusable rocket parts is a change in perspective that lines up with the standards of maintainability. By diminishing the requirement for the persistent creation of new rockets, reusability limits asset utilization, brings down send off costs, and adds to a more reasonable way to deal with space investigation.

Added substance Assembling:

Headways in added substance fabricating, ordinarily known as 3D printing, have altered the development of rocket parts. Customary assembling techniques include the get together of perplexing designs from various parts, frequently requiring complex machining and welding. Conversely, added substance fabricating takes into account the making of multifaceted and enhanced parts layer by layer, lessening burn through and creation time.

Rocket motors, parts, and, surprisingly, whole rocket stages can be made utilizing 3D printing procedures. This advancement gives more noteworthy plan adaptability and empowers the development of lightweight yet hearty designs. The diminished weight adds to expanded payload limit, working on the general proficiency of rocket dispatches.

The advantages of added substance producing reach out past proficiency to supportability. By limiting material waste and smoothing out the creation interaction, 3D printing lines up with the standards of earth cognizant assembling. Furthermore, the capacity to rapidly emphasize and test plans works with advancement, speeding up the improvement of new drive innovations.

Green Charges:

Ecological worries related with conventional rocket fuels, especially the arrival of poisons into the climate, have prodded investigation into greener other options. Green charges plan to limit the natural effect of room dispatches by lessening the poisonousness and destructive side-effects related with customary fuels.

One outstanding model is hydroxylammonium nitrate-based fuels. These definitions offer lower harmfulness than customary hypergolic fuels and have shown execution tantamount to regular charges. Examination into green charges lines up with more extensive endeavors to moderate the natural impression of room investigation.

NASA's Green Fuel Mixture Mission (GPIM) is a spearheading drive that plans to show the feasibility of green charges in space. The mission uses a shuttle outfitted with a green drive framework, exhibiting the natural advantages of these other options.

The reception of green fuels addresses a basic move toward more maintainable space investigation. As the recurrence of room missions increments, relieving the ecological effect of rocket dispatches becomes basic for the drawn out feasibility of room investigation tries.

Atomic Warm Impetus:

The quest for higher proficiency in drive has prompted the investigation of cutting edge ideas like atomic warm impetus (NTP). NTP use atomic reactors to warm a force, commonly hydrogen, to high temperatures. The fast exhaust created by the warmed charge brings about higher explicit motivation contrasted with conventional compound rockets.

NTP offers the potential for altogether decreased travel times for ran missions to objections like Mars. The expanded explicit drive permits shuttle to accomplish higher speeds, empowering quicker travels and limiting the openness of space explorers to the brutal space climate.

While NTP holds guarantee for interplanetary travel, specialized difficulties should be tended to before it turns into a reasonable reality. These difficulties incorporate the improvement of vigorous reactor frameworks, compelling radiation protecting, and the protected treatment of

atomic materials. In any case, the potential proficiency acquires make atomic warm impetus an alluring choice for future profound space investigation.

Imaginative Ideas:

Past laid out impetus advances, visionary ideas are pushing the limits of what is conceivable in space investigation. The EmDrive, an electromagnetic impetus framework, has produced critical interest and discussion inside established researchers. The EmDrive purportedly produces push without the requirement for fuel removal by making electromagnetic reverberation inside a tightened cavity.

Trial results have shown minute degrees of pushed, however the hidden material science of the EmDrive stay a subject of discussion. While certain specialists question the legitimacy of the noticed push, others are charmed by the capability of this flighty drive idea. Whenever demonstrated feasible, the EmDrive could open additional opportunities for space apparatus impetus, testing how we might interpret old style physical science.

Another hypothetical idea that charms the creative mind is the Alcubierre drive. Proposed by physicist Miguel Alcubierre, the Alcubierre drive includes making a twist bubble in spacetime, considering quicker than-light travel. While the idea is steady with general relativity, the commonsense acknowledgment of an Alcubierre drive faces critical difficulties, including the requirement for extraordinary types of issue with negative energy thickness.

These creative ideas feature the inventive and ground breaking nature of drive research. While certain ideas might stay speculative or face huge specialized obstacles, they highlight the persevering quest for novel plans to push humankind into the following wilderness of room investigation.

Global Coordinated effort:

In the mission for drive effectiveness and supportability, global co-operation has turned into a sign of room investigation endeavors. The Worldwide Space Station (ISS) remains as a demonstration of the collaboration between countries in propelling human spaceflight capacities and leading logical examination in a microgravity climate.

Coordinated effort stretches out past government space organizations to incorporate privately owned businesses and global associations. The Artemis program, drove by NASA, means to return people to the Moon and make ready for future manned missions to Mars. This aggressive drive includes joint effort with global accomplices, including the European Space Organization (ESA), the Canadian Space Office (CSA), and others.

SpaceX, a confidential American aviation producer, plays had a significant impact in driving development and joint effort in the business space area. The organization's Team Mythical beast rocket, created in association with NASA, has exhibited the suitability of business maintained spaceflight. The progress of Group Mythical beast missions to the ISS grandstands the potential for public-private organizations in propelling space investigation.

As countries and confidential elements cooperate on shared objectives, the pooling of assets, skill, and abilities speeds up progress in impetus advances. The cooperative methodology encourages advancement as well as reinforces the establishment for reasonable and effective space investigation attempts.

Moral and Administrative Contemplations:

As drive innovations advance, moral, legitimate, and administrative contemplations come to the front. The dependable utilization of room assets, the counteraction of room flotsam and jetsam, and the even-handed circulation of advantages from space investigation are essential parts of a reasonable spacefaring future.

The Space Settlement, laid out in 1967, gives a system to overseeing the exercises of countries in space. Key standards incorporate the quiet utilization of space, the anticipation of weaponization, and the evasion of unsafe pollution. As space exercises increment, the significance of adherence to peaceful accords and standards becomes central.

Moral contemplations additionally reach out to the possible results of room investigation on divine bodies and extraterrestrial conditions. The Planetary Assurance strategies laid out by space organizations mean to forestall defilement of other heavenly bodies with Earth organic entities as well as the other way around. These strategies guarantee that logical examinations of different planets and moons are led with negligible impedance from earthly life.

The dependable and moral utilization of impetus advancements includes limiting the age of room garbage. Orbital trash represents a gamble to dynamic satellites and space apparatus, and endeavors to address space flotsam and jetsam moderation are basic for supporting the drawn out ease of use of Earth's circles.

6

Chapter 5

Spacecraft and Satellites: Rocketry Applications

Shuttle and satellites address the zenith of human inventiveness and innovative ability, pushing the limits of investigation and correspondence. These wonders of designing have become basic instruments for logical exploration, public safety, and worldwide availability. At the core of these undertakings lies rocketry, a field that has developed fundamentally since its beginning.

The historical backdrop of rocketry can be followed back to old times, with the development of explosive in China around the ninth hundred years. In any case, it was only after the twentieth century that rocketry took a monster jump forward with the spearheading work of visionaries like Konstantin Tsiolkovsky, Robert H. Goddard, and Hermann Oberth. These early trailblazers laid the preparation for the advancement of space apparatus and satellites, making way for humankind's introduction to space.

The coming of The Second Great War denoted a critical defining moment for rocketry, as military powers perceived the essential significance of rocket innovation. The German V-2 rocket, created by Wernher von Braun, turned into the world's most memorable long-range directed long range rocket. After the conflict, von Braun and other German researchers were brought to the US as a component of Activity Paperclip, adding to the expanding space race between the US and the Soviet Association.

The Space Age formally started with the send off of the Soviet satellite Sputnik 1 on October 4, 1957. This b-ball estimated orbiter not just proclaimed the beginning of human space investigation yet in addition started the space race between the superpowers. The US answered with the send off of Adventurer 1 out of 1958, denoting its entrance into the space race and the foundation of NASA, the Public Flying and Space Organization.

As the space race unfurled, the improvement of progressively strong rockets turned into a key concentration. The Saturn V, created by NASA for the Apollo program, stays the most remarkable rocket at any point constructed. Remaining more than 36 stories tall, the Saturn V assumed a critical part in sending people to the Moon interestingly during the memorable Apollo 11 mission in 1969.

Past human spaceflight, the rise of satellites opened up new boondocks in correspondence, Earth perception, and logical examination. Satellites can be extensively ordered into two sorts: normal satellites, like the Moon, and fake satellites, which are human-made objects set into space around heavenly bodies. The last option has turned into a foundation of present day innovation and framework.

Correspondence satellites, specifically, have altered worldwide network. Geostationary satellites, situated in circle over the equator and pivoting at similar speed as the Earth, empower consistent correspondence with fixed ground areas. This has worked with the development of media communications, TV broadcasting, and internet providers, associating individuals all over the planet in manners impossible only years and years prior.

Earth perception satellites, outfitted with cutting edge sensors and imaging innovation, give important information to checking the planet's environment, atmospheric conditions, and catastrophic events.

These satellites add to logical examination, ecological checking, and fiasco the executives, assuming a vital part in getting it and relieving the effects of environmental change.

Route satellites have become basic to present day route frameworks, empowering exact situating and timing for a horde of utilizations, from vehicle route frameworks to sea tasks. The Worldwide Situating Framework (GPS), worked by the US, was quite possibly the earliest and stays one of the most broadly utilized satellite route frameworks universally.

Logical exploration directed in space has given novel experiences into the universe past Earth. Telescopes like the Hubble Space Telescope, situated over Earth's climate, have caught amazing pictures of far off systems and nebulae, extending how we might interpret the universe. Space-based observatories keep on pushing the limits of galactic exploration, disclosing the secrets of dull matter, dim energy, and the starting points of the universe.

Progressions in scaling down and savvy satellite advances have prompted the ascent of little satellites, including CubeSats. These little, block molded satellites are frequently utilized for logical exploration, innovation exhibit, and instructive purposes. Their reduced size and moderately minimal expense have democratized admittance to space, permitting colleges, research organizations, and, surprisingly, privately owned businesses to take part in space investigation.

Lately, the confidential area has assumed an undeniably noticeable part in space investigation. Organizations like SpaceX, established by Elon Musk, have created reusable rocket innovation, essentially lessening the expense of sending off payloads into space. SpaceX's Hawk 9 rocket, fit for conveying both maintained and uncrewed payloads, has turned into a workhorse for satellite send-offs and resupply missions to the Global Space Station (ISS).

The ISS itself remains as a demonstration of worldwide cooperation in space investigation. A joint task including space organizations from the US, Russia, Europe, Japan, and Canada, the ISS fills in as a microgravity research lab and a stage for logical trials directed in the one of a kind climate of room. It likewise represents the potential for serene collaboration in the investigation and usage of space.

Looking forward, the fate of room investigation holds energizing prospects. Aggressive missions to Mars, for example, NASA's Artemis program and SpaceX's Starship project, intend to additionally extend human presence past Earth. The idea of room the travel industry is likewise getting some forward movement, with organizations imagining a future where confidential people can encounter the excitement of room travel.

Headways in impetus advancements, like particle drive and atomic impetus, guarantee quicker and more productive travel inside our planetary group and then some. These advancements could alter interplanetary investigation and prepare for manned missions to far off objections, like the moons of Jupiter and Saturn.

The quest for extraterrestrial life stays an enrapturing focal point of logical request. Missions to frosty moons like Europa (moon of Jupiter) and Enceladus (moon of Saturn), with their subsurface seas, hold the possibility to find indications of something going on under the surface past Earth. The investigation of exoplanets, planets circling stars outside our planetary group, keeps on uncovering a different cluster of universes, some of which might hold onto the circumstances fundamental for life as far as we might be concerned.

Space trash, involved outdated satellites, spent rocket stages, and parts from crashes, represents a developing test to the supportability of room exercises. Endeavors to resolve this issue incorporate the advancement of rules for capable space direct, the advancement of flotsam and jetsam relief measures, and the investigation of innovations to eliminate space garbage from circle effectively.

All in all, the field of rocketry has impelled humankind into the immensity of room, empowering the investigation and usage of the universe. Rocket and satellites, conceived out of the fantasies and assurance of visionaries, have become essential instruments for logical disclosure, correspondence, and Earth perception. As we stand near the precarious edge of another time in space investigation, filled by mechanical development and global coordinated effort, the opportunities for humankind's future in space are essentially as vast as the actual universe.

5.1 The role of rockets in launching satellites and space probes

Rockets have for some time been the doorway to the universe, driving satellites and space tests past Earth's environment to investigate the profundities of room. The job of rockets in sending off these pivotal payloads is complex, requiring accuracy, power, and unwavering quality. From the beginning of room investigation to the present, rockets have been the workhorses of the space business, working with logical disclosure, correspondence, and mechanical headways.

The basic guideline behind rocketry lies in Newton's third law of movement: for each activity, there is an equivalent and inverse response. Rockets work on the rule of removing mass at high velocities to produce push the other way. This push moves the rocket forward, defeating Earth's gravitational force and permitting it to arrive at the vacuum of room. The historical backdrop of rocketry is a demonstration of humankind's mission to tackle this standard for investigation and logical undertakings.

The origin of current rocketry can be ascribed to visionaries like Konstantin Tsiolkovsky, Robert H. Goddard, and Hermann Oberth. Tsiolkovsky, a Russian researcher, is in many cases viewed as the dad of astronautics. In the mid twentieth hundred years, he figured out the rocket condition, laying the hypothetical foundation for space travel. Simultaneously, in the US, Goddard was directing spearheading tests in fluid powered rockets. His work finished in the send off of the world's most memorable fluid filled rocket on Walk 16, 1926, in Reddish, Massachusetts. In the mean time, Oberth, a German researcher, distributed powerful deals with rocketry, adding to the comprehension of rocket drive.

WWII assumed a urgent part in propelling rocket innovation, with the improvement of long range rockets, for example, the German V-2 rocket. Wernher von Braun, a critical figure in the V-2 program, would later become instrumental in the US's space program. After the conflict, as a feature of Activity Paperclip, von Braun and other German researchers were brought to the U.S., where their skill established the groundwork for the nation's rocket improvement endeavors.

The Virus War contention between the US and the Soviet Association powered the space race, denoting another time in rocketry. The Soviet Association accomplished a notable achievement with the send off of Sputnik 1, the world's most memorable counterfeit satellite, on October 4, 1957.

The occasion denoted the beginning of the space age and incited the US to speed up its space program. The effective send off of Pilgrim 1 out of 1958 denoted America's entrance into space investigation, laying the preparation for the foundation of NASA.

As the space race unfurled, rockets turned out to be progressively strong, equipped for conveying heavier payloads into space. The improvement of the Saturn group of rockets by NASA, strikingly the Saturn V, assumed a significant part in the Apollo program. The Saturn V remaining parts the most remarkable rocket at any point constructed, fit for conveying space travelers to the Moon. The famous picture of the Saturn V taking off from Kennedy Space Center represents the stupendous accomplishment of the Apollo missions.

Satellites, both regular and fake, play had a significant impact in growing comprehension we might interpret the universe and upgrading life on The planet. Counterfeit satellites, specifically, have become indispensable to current life, with applications going from media communications and weather conditions determining to route and logical examination.

Correspondence satellites, situated in geostationary circle over the equator, empower worldwide network by working with TV broadcasting, internet providers, and significant distance correspondence. These satellites act as correspondence transfers, sending signals between ground stations and giving consistent correspondence administrations to remote and out of reach areas.

Earth perception satellites have reformed our capacity to screen and grasp the planet's elements. Furnished with cutting edge sensors and imaging innovation, these satellites give important information to weather conditions estimating, ecological observing, fiasco the executives, and logical exploration. Satellites, for example, Landsat have recorded changes in Earth's surface over many years, adding to how we might interpret environmental change and deforestation.

Route satellites, some portion of worldwide route satellite frameworks (GNSS), have changed the manner in which we explore the world. The Worldwide Situating Framework (GPS), created by the US, is a heavenly body of satellites that empowers exact situating and timing data. GPS is vital to applications going from route frameworks in vehicles and cell phones to aeronautics and oceanic activities.

Logical satellites and space tests have broadened mankind's venture into the universe. Telescopes like the Hubble Space Telescope, situated over Earth's environment, have given exceptional perspectives on far off systems, nebulae, and other divine peculiarities. Automated space tests, for example, the Explorer tests and the Mars meanderers, have investigated the external compasses of our planetary group and the outer layer of different planets, opening the secrets of the universe.

The effective send off of satellites and space tests relies on the abilities of the rockets that convey them into space. The decision of rocket relies upon different variables, including the payload's mass, objective, and required circle. Various sorts of rockets, for example, disposable send off vehicles and reusable send off vehicles, offer unmistakable benefits and compromises.

Superfluous send off vehicles, encapsulated by rockets like the Map book V and Delta IV, are intended for a solitary use. When they convey their payload to circle, the rocket's parts are disposed of. These rockets are appropriate for sending off weighty payloads into geostationary circle or sending logical tests on directions past Earth's circle. In any case, the expense of assembling new rockets for each send off can be restrictively high.

Conversely, reusable send off vehicles, exemplified by SpaceX's Bird of prey 9 and Hawk Weighty, are intended to get back to Earth and be reused for various send-offs. This reusability fundamentally decreases send off costs, opening up additional opportunities for more successive and financially savvy admittance to space. The main phase of a reusable rocket, in the wake of confining from the payload, executes a controlled plunge and landing, fit to be repaired and flown once more.

Headways in rocket drive have likewise assumed a basic part in improving send off capacities. Conventional synthetic rockets, utilizing fluid or strong charges, stay the prevailing drive innovation. In any case, advancements, for example, particle drive, which ousts particles at high rates to create push, offer expanded proficiency and efficiency for profound space missions. Atomic drive, an idea under investigation, could give much more noteworthy push to manned missions to far off objections.

The improvement of satellite send off vehicles includes careful preparation and designing to guarantee an effective mission. The payload's mass and expected circle direct the rocket's size and design. Send off locales are decisively situated to exploit Earth's pivot and guarantee ideal directions for arriving at explicit circles.

The send off succession itself is a painstakingly organized series of occasions. The rocket goes through a cycle known as commencement, during which frameworks are checked, fuels are stacked, and last arrangements are made. When the commencement arrives at nothing, the rocket motors touch off, and the vehicle starts its climb into space. The rocket goes through various stages, each containing motors and fuel tanks. As each stage debilitates its force, it is casted off to decrease weight and permit the excess stages to speed up the payload to its expected speed.

The organization of satellites into their assigned circles is a basic period of the send off. Contingent upon the mission prerequisites, the rocket's upper stage executes a progression of moves to deliver the payload into a particular circle.

For satellites bound for geostationary circle, the upper stage might play out a circularization consume to accomplish the necessary elevation and tendency.

Once in circle, satellites and space tests depend on locally available drive frameworks for orbital moves, station-keeping, and deorbiting toward the finish of their functional life. Little engines or particle impetus frameworks are normally utilized for exact orbital changes. A few satellites, furnished with sun based cruises or high level drive, can embrace stretched out missions to investigate various circles or travel past our planetary group.

The idea of room garbage, made out of dead satellites, spent rocket stages, and parts from impacts, represents a developing worry for the supportability of room exercises. As the volume of room garbage increments, so does the gamble of impacts that could produce more flotsam and jetsam in a flowing impact known as the Kessler disorder. Endeavors to moderate space trash incorporate rules for mindful space lead, the advancement of flotsam and jetsam evacuation innovations, and worldwide coordinated effort to address this common test.

The job of rockets in space investigation stretches out past satellite send-offs to envelop human spaceflight. Manned missions to the Worldwide Space Station (ISS) and aggressive designs for human investigation of the Moon and Mars depend on strong rockets to move space explorers and their hardware into space. The Space Send off Framework (SLS), created by NASA, addresses the up and coming age of weighty lift rockets intended for human space investigation.

Looking forward, the eventual fate of rocketry holds energizing prospects and difficulties. Privately owned businesses, driven by enterprising visionaries, are assuming an undeniably unmistakable part in space investigation. SpaceX, established by Elon Musk, has exhibited the reasonability of reusable rockets and plans to make humankind a multiplanetary animal varieties with the improvement of the Starship shuttle. Different organizations, like Blue Beginning and Rocket Lab, are additionally adding to the advancement of rocket innovation.

Notwithstanding progressions in drive and send off innovations, worldwide coordinated effort is turning into a sign of room investigation. Cooperative endeavors, like the European Space Organization (ESA) and joint missions like the James Webb Space Telescope (JWST), feature the common obligation to unwinding the secrets of the universe.

The investigation of room, whether through satellites, space tests, or maintained missions, keeps on enthralling the human creative mind. It pushes the limits of what is conceivable and moves new ages of researchers, specialists, and adventurers. As we stand on the limit of remarkable accomplishments in space investigation, the job of rockets stays fundamental to opening the secrets of the universe and graphing the course for mankind's future in the huge spread of room.

5.2 Overview of iconic spacecraft and their contributions to space exploration

Space investigation has been set apart by the creation and sending of famous shuttle that have altogether progressed how we might interpret the universe and extended the outskirts of human information. These shuttle, going from mechanical tests to ran vehicles, play played vital parts in logical disclosure, innovative advancement, and global joint effort. This outline digs into the absolute most notorious shuttle ever, featuring their commitments to space investigation.

Perhaps of the earliest achievement in space investigation was accomplished with the send off of the Soviet satellite Sputnik 1 on October 4, 1957. Sputnik 1 denoted the beginning of the space age, turning into the world's most memorable counterfeit satellite. This b-ball measured orbiter, furnished with radio transmitters, gave significant information on the thickness of Earth's upper climate and showed the achievability of putting objects into Earth's circle. Sputnik 1's prosperity prepared for ensuing space missions and lighted the space race between the US and the Soviet Association.

Following the progress of Sputnik 1, the Soviet Association kept on gaining ground in space investigation with the send off of Yuri Gagarin, the principal human in space, on board the Vostok 1 space apparatus on April 12, 1961. Gagarin's noteworthy circle around Earth denoted a critical accomplishment in maintained spaceflight and hardened the Soviet Association's lead in the early long periods of the space race.

In light of the Soviet Association's triumphs, the US sloped up its space program, prompting the foundation of NASA (Public Aviation and Space Organization) in 1958. One of NASA's initial accomplishments was the Mercury program, which intended to place American space explorers into space. The Mercury shuttle, including the famous Opportunity 7 container that conveyed Alan Shepard on the principal American human spaceflight in 1961, established the groundwork for ensuing ran missions.

The Gemini program, which followed Mercury, further high level American space capacities. The Gemini shuttle, intended for two space explorers, empowered crucial moves like orbital meeting and docking, laying the preparation for future Moon missions. Gemini 4, directed by James McDivitt and steered by Ed White, accomplished the principal American spacewalk in 1965, an achievement in extravehicular action (EVA) that made ready for future spacewalks.

The Apollo program, the lead of American maintained space investigation, intended to land space travelers on the Moon and take them securely back to Earth. The Apollo shuttle comprised of the Order Module, where space explorers resided and worked, the Help Module, giving impetus and emotionally supportive networks, and the Lunar Module, intended for lunar plummet and rising.

The notable Apollo 11 mission, with space explorers Neil Armstrong, Buzz Aldrin, and Michael Collins, effectively handled the main people on the Moon on July 20, 1969. Armstrong's notorious words, "That is one little step for [a] man, one monster jump for humankind," denoted a great accomplishment in mankind's set of experiences.

The Space Transport program, started by NASA in 1981, presented another time in space transportation with reusable rocket. The Space Transport comprised of an orbiter, strong rocket supporters, and an outer gas tank. The orbiter, like the notable Atlantis, Columbia,

Disclosure, Attempt, and Challenger, conveyed space explorers and payloads into space. The Space Transport assumed a pivotal part in sending satellites, directing logical examinations, and gathering and overhauling space stations, for example, Skylab and the Global Space Station (ISS).

The Hubble Space Telescope, sent off on board the Space Transport Disclosure in 1990, became one of the most famous and powerful space observatories. Situated over Earth's climate, Hubble gave exceptional perspectives on far off worlds, nebulae, and other divine items. Its high-goal pictures and important information have contributed altogether to how we might interpret the universe, from the estimation of the pace of development of the universe to the revelation of exoplanets.

The Global Space Station (ISS), a cooperative exertion including different space organizations, has been an image of worldwide participation in space. The ISS fills in as a microgravity research lab and a stage for logical examinations in the novel climate of room. Modules from different nations, sent off and gathered in space, structure the design of the ISS. Manned missions on board rocket like the Russian Soyuz, American SpaceX Group Mythical serpent, and Boeing CST-100 Starliner add to the continuous examination and trial and error on the ISS.

Automated shuttle have likewise assumed a pivotal part in space investigation, wandering past Earth to investigate the planets, moons, and other heavenly collections of our planetary group. The Explorer tests, sent off in 1977, set out on a fabulous visit through the external planets — Jupiter, Saturn, Uranus, and Neptune. Explorer 1 and Explorer 2 gave important information and pictures, including the notorious "Light Blue Dab" photograph taken by Explorer 1 from the edge of the planetary group, displaying Earth as a small bit in the boundlessness of room.

The Mars Meanderer missions, including Sojourner, Soul, Opportunity, Interest, and Persistence, have been instrumental in the investigation of the Martian surface. These meanderers, furnished with cutting edge logical instruments, have given significant experiences into the topography, environment, and likely tenability of Mars. The notable pictures of Martian scenes caught by these meanderers have enraptured the public's creative mind and energized interest in future maintained missions to the Red Planet.

The Cassini-Huygens mission, a cooperation between NASA, the European Space Office (ESA), and the Italian Space Organization (ASI), investigated the Saturn framework. Sent off in 1997, the Cassini orbiter concentrated on Saturn, its rings, and its moons for north of 13 years. The Huygens test, delivered by Cassini, dropped to the outer layer of Saturn's biggest moon, Titan, giving important information on its environment and surface circumstances.

The New Skylines mission, sent off in 2006, led a noteworthy flyby of Pluto in 2015, giving the primary close-up pictures and information of the far off bantam planet. New Skylines proceeded with its excursion into the Kuiper Belt, concentrating on other little, cold bodies in the external ranges of our planetary group. The mission extended how we might interpret the different universes past the circle of Neptune.

As of late, privately owned businesses have arisen as central participants in space investigation, creating imaginative space apparatus and send off vehicles. SpaceX, established by Elon Musk, has been at the front of this new time. The Hawk 9 and Bird of prey Weighty rockets, alongside the Group Mythical serpent space apparatus, have shown reusable rocket innovation and shipped space explorers to the ISS. SpaceX's Starship, intended for maintained missions to the Moon, Mars, and then some, addresses an intense vision for the eventual fate of room investigation.

Blue Beginning, established by Jeff Bezos, is another privately owned business zeroing in on space investigation. The New Shepard suborbital rocket and shuttle mean to empower business space the travel industry, permitting private people to encounter brief times of weightlessness in space. Blue Beginning's New Glenn rocket, presently being developed, is intended for orbital missions and satellite send-offs.

As space investigation keeps on advancing, new rocket and missions are being wanted to investigate the Moon, Mars, and different objections in our nearby planet group and then some. The Artemis program, drove by NASA, plans to return people to the lunar surface and lay out a maintainable human presence on the Moon. Plans for manned missions to Mars are likewise being effectively investigated by both government and confidential elements, imagining a future where people become an interplanetary species.

5.3 Satellite technology and its impact on modern life

Satellite innovation remains as a foundation of present day life, molding and impacting different parts of our everyday presence in manners frequently underestimated. From correspondence and route to weather conditions guaging and logical exploration, satellites have become key devices that interface the world and give significant data to a bunch of utilizations. This outline digs into the development of satellite innovation and its significant effect on present day life.

The starting points of satellite innovation can be followed back to the mid-twentieth hundred years, a period set apart by the beginning of the space age and the investigation of space. The send off of the principal counterfeit satellite, Sputnik 1, by the Soviet Association in 1957, denoted a urgent crossroads ever. This ball estimated orbiter, outfitted with radio transmitters, not just shown the achievability of setting objects into Earth's circle yet in addition established the groundwork for the usage of satellites in different fields.

Correspondence satellites address one of the earliest and most critical utilizations of satellite innovation. Geostationary correspondence satellites, situated in circle over the equator, have reformed worldwide availability. These satellites, like the Intelsat series, work with media communications, TV broadcasting, and internet providers by transferring signals between ground stations and giving steady inclusion to explicit areas.

The idea of geostationary circle, where a satellite circles the Earth at a similar rotational speed as the planet, permits these correspondence satellites to stay fixed comparative with a particular point on the World's surface. This trademark empowers ceaseless correspondence with fixed ground areas, causing them ideal for administrations that to require consistent availability, for example, TV broadcasting and web correspondence.

Direct transmission satellites, similar to those utilized for satellite TV administrations, work in lower Earth circles (LEO) and give direct-to-home telecom. These satellites, remembering those for the DIRECTV and Dish Organization star groupings, bar TV flags straightforwardly to little dish recieving wires on the ground, offering a large number of stations and programming choices.

The effect of correspondence satellites on present day life is significant, stretching out to the domains of worldwide media communications and worldwide network. Undersea fiber optic links might deal with most of worldwide web traffic, however satellites assume a significant part in interfacing remote and secluded districts where it is unfeasible or cost-restrictive to lay links. Satellites have crossed over correspondence holes, connecting individuals across mainlands and furnishing fundamental network in regions with restricted earthly framework.

Route satellites address one more extraordinary use of satellite innovation. Worldwide Route Satellite Frameworks (GNSS), like the Worldwide Situating Framework (GPS), have become basic to current route and situating. GPS, worked by the US government, comprises of a group of stars of satellites in medium Earth circle (MEO) that give exact situating, route, and timing data to clients around the world.

The universality of GPS innovation has upset route in different spaces, from individual gadgets like cell phones and vehicle route frameworks to aeronautics, oceanic tasks, and crisis administrations.

Exact situating data empowers proficient course arranging, continuous following, and area based administrations, adding to expanded security, accommodation, and productivity in various parts of day to day existence.

Logical exploration and Earth perception comprise one more fundamental domain of satellite innovation. Earth perception satellites, furnished with cutting edge sensors and imaging innovation, give priceless information to observing and grasping the planet's elements. These satellites, like those in the Landsat series, add to ecological checking, asset the executives, and debacle reaction.

Satellites assume a critical part in weather conditions guaging and environment research. Meteorological satellites, remembering those for the Geostationary Functional Natural Satellite (GOES) series, constantly screen the World's environment, giving continuous information on weather conditions, overcast cover, and climatic circumstances. This data is instrumental in anticipating and following storms, hurricanes, and other extreme climate occasions, empowering convenient alerts and debacle readiness.

Remote detecting satellites, outfitted with particular sensors, catch high-goal pictures and information for applications going from agribusiness and ranger service to metropolitan preparation and natural protection. These satellites, exemplified by the European Space Organization's Copernicus program and the Sentinel satellites, add to checking deforestation, evaluating land use changes, and giving fundamental information to reasonable asset the executives.

Satellites have likewise assumed a critical part in logical investigation past Earth. Space telescopes, situated over Earth's climate to stay away from environmental obstruction, catch shocking pictures and gather information on far off divine articles. The Hubble Space Telescope, sent off in 1990, has been instrumental in propelling comprehension we might interpret the universe, catching definite pictures of worlds, nebulae, and other cosmic peculiarities.

Route in space, particularly for interplanetary missions, depends on the utilization of satellites and space tests. Space organizations, including NASA and the European Space Office, have sent a progression of mechanical shuttle to investigate the planets, moons, and space rocks inside our planetary group. These space tests, like the Mars meanderers and the Juno mission to Jupiter, send important information back to Earth, growing our insight into the universe.

Satellites have additionally become pivotal devices in the field of protection and public safety. Military correspondence satellites, similar to those worked by the US Branch of Guard, empower secure and dependable correspondence for safeguard powers all over the planet. Surveillance satellites give high-goal pictures and knowledge, supporting military tasks and vital preparation.

The improvement of satellite innovation has not been restricted to huge, government-drove programs. Lately, progressions in scaling down and savvy satellite plan have prompted the ascent of little satellites, including CubeSats. These little, block molded satellites are frequently utilized for logical examination, innovation show, and instructive purposes. Their conservative size and generally minimal expense have democratized admittance to space, permitting colleges, research establishments, and, surprisingly, privately owned businesses to partake in space investigation.

The development of the confidential area as a key part in space investigation has reshaped the scene of satellite innovation. Organizations like SpaceX, established by Elon Musk, have created imaginative answers for satellite send-offs and space transportation. SpaceX's Hawk 9 rocket, with its reusable first stage, has essentially diminished the expense of sending off payloads into space, making it an appealing choice for satellite organization.

Perhaps of the most aggressive venture in satellite innovation is the advancement of megaconstellations — enormous organizations of interconnected satellites that cooperate to give worldwide inclusion. Organizations like SpaceX (Starlink), OneWeb, and Amazon (Task Kuiper) are chipping away at sending megaconstellations to convey rapid web admittance to underserved and far off regions all over the planet. These star groupings intend to connect the advanced separation and carry dependable web availability to locales where customary framework is inadequate.

While the advantages of satellite innovation are evident, difficulties and contemplations additionally exist. Space flotsam and jetsam, contained old satellites, spent rocket stages, and pieces from crashes, represents a developing danger to the manageability of room exercises. Endeavors to resolve this issue incorporate the advancement of rules for capable space lead, the advancement of flotsam and jetsam moderation measures, and the investigation of innovations to eliminate space garbage from circle effectively.

The rising blockage in Earth's circle, driven by the developing number of satellites and space exercises, has provoked conversations about space traffic the board and the requirement for worldwide collaboration to guarantee the capable utilization of room. Administrative systems and rules are being created to resolve issues like crash evasion, orbital trash relief, and the drawn out maintainability of room exercises.

Looking forward, the fate of satellite innovation holds invigorating conceivable outcomes and difficulties. The arrangement of cutting edge sensors, computerized reasoning, and AI in satellites will upgrade their capacities for Earth perception, natural observing, and logical examination. Inventive impetus advancements, like particle drive and sun oriented sails, offer the potential for additional productive and practical satellite missions.

Progressions in satellite correspondence innovations, including high-throughput satellites and high level sign handling, will add to further developed network and information transmission abilities. As mega-constellations become functional, they can possibly reshape the scene of worldwide web access, giving fast availability to remote and under-served regions.

Interplanetary investigation will go on with the send off of new missions to concentrate on Mars, the Moon, and other heavenly bodies. The improvement of manned missions to the Moon, Mars, and past will depend on satellite innovation for route, correspondence, and Earth perception. Privately owned businesses will keep on assuming a conspicuous part, driving development and contest in the space.

The appearance of innovation has changed basically every part of current life, impacting the manner in which we live, work, impart, and access data. The effect of innovation is unavoidable, shaping social orders, economies, and individual encounters in remarkable ways. This outline dives into the complex effect of innovation on current life, investigating its consequences for correspondence, schooling, medical services, work, and social collaborations.

Correspondence remains as perhaps of the most significantly impacted space by mechanical progressions. The development of correspondence advancements, from the message to the web, has fundamentally adjusted the manner in which individuals associate with one another across the globe. The ascent of the web and cell phones has prompted immediate and pervasive correspondence, permitting people to associate progressively paying little mind to geological distances.

Virtual entertainment stages, like Facebook, Twitter, Instagram, and LinkedIn, have become basic to current correspondence, empowering individuals to share encounters, sentiments, and updates with a worldwide crowd. These stages work with the arrangement of virtual networks, where people can associate in light of shared interests, encounters, or expert affiliations. The instantaneousness and openness of online entertainment have reshaped how news is scattered, occasions are reported, and social developments pick up speed.

Notwithstanding relational correspondence, innovation has reformed how data is gotten to and consumed. The web fills in as a far reaching vault of information, giving moment admittance to an abundance of data on basically any subject. Web search tools like Google have become key instruments for recovering data, while online stages, for example, Wikipedia offer cooperative and continually refreshed wellsprings of information.

The multiplication of cell phones has additionally intensified the effect of innovation on correspondence and data access. Portable applications, or applications, take care of many necessities, from interpersonal interaction and news utilization to efficiency and amusement. The universality of cell phones has made data and correspondence more customized, permitting people to fit their internet based encounters to their inclinations and interests.

Schooling has gone through a significant change because of innovative progressions. The reconciliation of innovation in schooling, frequently alluded to as EdTech, has disturbed conventional training techniques and extended admittance to learning open doors. Web based learning stages, like Coursera, edX, and Khan Foundation, offer courses and assets that rise above geological hindrances, giving instruction to people around the world.

The utilization of innovation in homerooms, from intuitive whiteboards to instructive programming, improves the growth opportunity by making it more intuitive, drawing in, and custom-made to individual requirements. Virtual and expanded reality advances have presented vivid opportunities for growth, permitting understudies to investigate verifiable destinations, direct virtual examinations, and draw in with instructive substance in extraordinary ways.

E-learning and distant training acquired critical unmistakable quality during the Coronavirus pandemic, featuring the flexibility and versatility of innovation in guaranteeing progression in schooling. Virtual homerooms, video conferencing instruments, and cooperative web-based stages became fundamental parts of remote getting the hang of, empowering understudies to get to instruction from the wellbeing of their homes.

Medical care has likewise seen an extraordinary effect from innovation, prompting progressions in clinical finding, therapy, and patient consideration. The digitization of clinical records and the improvement of electronic wellbeing record (EHR) frameworks have smoothed out the administration and openness of patient data, upgrading coordination among medical care suppliers and working on quiet results.

Telemedicine, worked with by video conferencing and remote observing innovations, has arisen as a vital part of current medical services conveyance. Telehealth administrations empower patients to talk with medical care experts, get clinical exhortation, and even go through virtual assessments from the solace of their homes. This has demonstrated especially significant in remote or underserved regions, where admittance to medical services offices might be restricted.

Clinical imaging innovations, for example, attractive reverberation imaging (X-ray) and figured tomography (CT) filters, have progressed with innovation, giving more definite and exact symptomatic data. Wearable gadgets and wellbeing following applications permit people to screen their wellbeing measurements, advancing preventive consideration and working with early mediation.

The field of hereditary qualities has encountered an upset with the coming of genomics and quality altering innovations. The Human Genome Venture, finished in 2003, prepared for figuring out the human hereditary code, opening bits of knowledge into hereditary inclinations for illnesses and working with customized medication.

CRISPR-Cas9, a progressive quality altering instrument, holds the commitment of designated hereditary changes for treating hereditary problems.

In the domain of work, innovation has reshaped the idea of business, the construction of associations, and how undertakings are performed. The ascent of computerization, man-made reasoning (artificial intelligence), and AI has prompted the robotization of normal and redundant assignments across different enterprises. While computerization can possibly further develop proficiency and lessen physical work, it likewise brings up issues about the effect on business and the requirement for reskilling and upskilling.

Remote work, worked with by correspondence and joint effort innovations, acquired noticeable quality, particularly during the Coronavirus pandemic. Virtual gatherings, cloud-based cooperation devices, and task the executives programming permitted representatives to keep up with efficiency while telecommuting. This change in work elements provoked a reconsideration of customary office-based models, with numerous associations embracing crossover work models that mix remote and in-person work.

The gig economy, empowered by computerized stages and portable applications, has arisen as an unmistakable component of the cutting edge work scene. People can take part in present moment, independent, or provisional labor worked with by stages like Uber, Lyft, TaskRabbit, and Upwork. While the gig economy gives adaptability to laborers, it additionally raises worries about employer stability, work freedoms, and the shortfall of customary business benefits.

Mechanical headways have additionally changed the scene of diversion and media utilization. Web-based features, like Netflix, Hulu, and Spotify, offer on-request admittance to an immense range of films, Programs, music, and digital recordings. The shift from conventional link and satellite TV to streaming stages has given shoppers more noteworthy command over when, where, and how they consume content.

Advanced distributing and digital books have upset the distributing business, giving writers and perusers new roads for making and getting to composed content. Online entertainment stages have become compelling channels for content creation, circulation, and utilization, permitting people and associations to contact wide crowds and draw in with different networks.

Social associations have been significantly affected by innovation, especially through the ascent of web-based entertainment and advanced correspondence stages. Long range informal communication locales, informing applications, and online discussions have become vital to how individuals associate, share encounters, and structure networks. In any case, the effect of web-based entertainment on emotional well-being, protection concerns, and the spread of deception are likewise subjects of continuous talk and exploration.

The union of innovation with transportation has prompted critical advancements in the manner individuals move and drive. Ride-sharing administrations, like Uber and Lyft, have changed metropolitan transportation by giving helpful and adaptable options in contrast to conventional cabs. Electric vehicles, independent vehicles, and progressions in open transportation frameworks add to the continuous development of the portability scene.

The Web of Things (IoT) has interconnected ordinary gadgets, from brilliant indoor regulators and wearable wellness trackers to home security frameworks and associated kitchen machines. These interconnected gadgets empower the assortment and trade of information, setting out open doors for computerization, productivity, and improved client encounters. In any case, the multiplication of IoT gadgets additionally raises worries about information protection and security.

The effect of innovation reaches out past individual areas to cultural and worldwide difficulties. Environmental change, supportability, and ecological preservation are regions where innovation is assuming a vital part. Developments in sustainable power innovations, shrewd framework frameworks, and manageable horticulture rehearses add to tending to natural worries and building a more practical future.

Online protection has turned into a basic part of the cutting edge computerized scene. The rising dependence on advanced frameworks and availability carries with it the gamble of digital dangers, going from information breaks and ransomware assaults to wholesale fraud. The requirement for powerful network safety measures and the advancement of secure innovations are basic to safeguard people, associations, and basic foundation.

Moral contemplations encompassing innovation, including issues connected with information security, algorithmic predisposition, and the mindful advancement of artificial intelligence, stand out enough to be noticed. Conversations about computerized morals, dependable man-made intelligence arrangement, and the moral utilization of arising innovations are vital to guaranteeing that mechanical progressions line up with cultural qualities and standards.

Chapter 6

Challenges and Triumphs: The International Space Station (ISS)

The Global Space Station (ISS) remains as a demonstration of human creativity, worldwide joint effort, and the steady soul of investigation. Since its origin, the ISS has been an image of mankind's capacity to conquer imposing difficulties and accomplish momentous victories in the unfriendly climate of room. From its origination to its continuous tasks, the ISS has explored a perplexing trap of specialized, political, and calculated difficulties, arising as a guide of logical accomplishment and worldwide participation.

The starting points of the ISS follow back to the beginning of room investigation, when both the US and the Soviet Association imagined the chance of a space station. Notwithstanding, it was the finish of the Virus War that made ready for extraordinary coordinated effort between previous adversaries. The organization among NASA and the Russian space organization, Roscosmos, established the groundwork for the ISS, changing it from a Virus War vision into an image of solidarity among countries.

The principal significant test in building the ISS was fostering the innovation and designing arrangements expected to collect and keep a tenable construction in the brutal climate of low Earth circle. This try requested the mix of different innovations, from life emotionally supportive networks to cutting edge advanced mechanics. The complexities of gathering the ISS piece by piece in space introduced a variety of specialized difficulties, with every module requiring exact coordination and careful preparation.

One of the vital victories in beating these difficulties was the effective send off and get together of the Russian Zarya module in 1998. This obvious the start of a persistent human presence in space and set up for ensuing module increases. The measured way to deal with development considered the consolidation of parts from different worldwide accomplices, including the European Space Organization (ESA), the Japan Aviation Investigation Organization (JAXA), and the Canadian Space Organization (CSA).

The intricacy of overseeing such a different and global venture presented another arrangement of difficulties: organizing endeavors among countries with various dialects, societies, and specialized principles. The victory in such manner was the foundation of a bound together global structure that empowered consistent coordinated effort. The ISS turned into a model for quiet participation, showing the way that even countries with verifiable contrasts could cooperate to benefit logical investigation.

As the ISS extended, so did the difficulties related with keeping up with its functional abilities. The brutal states of room, including microgravity, outrageous temperatures, and openness to vast radiation, presented continuous dangers to the station's primary trustworthiness and the prosperity of its team. Wins in innovation, for example, high level materials and imaginative designing arrangements, assumed a urgent part in beating these difficulties.

The consistent presence of people in space likewise introduced physiological difficulties. Expanded times of weightlessness can significantly affect the human body, including muscle decay, bone thickness misfortune, and changes in cardiovascular capability. The victory here lay in the advancement of countermeasures and exercise systems that permitted space explorers to alleviate the adverse impacts of delayed spaceflight, empowering them to live and work in space for broadened terms.

Logical examination on board the ISS has yielded an abundance of information across different disciplines, from principal material science to science and medication. Be that as it may, directing analyses in microgravity presented its own arrangement of difficulties. Analysts needed to adjust their philosophies to represent the shortfall of gravity, and architects needed to configuration specific gear to work in the novel climate of room. The victory, for this situation, was the age of noteworthy logical revelations that have sweeping ramifications for both space investigation and life on The planet.

The ISS has likewise been a proving ground for state of the art innovations that can possibly upset space investigation. Mechanical technology and independent frameworks play had a pivotal impact in the get together and support of the station. The Canadarm2 mechanical arm, created by the Canadian Space Organization, has been instrumental in catching and berthing visiting rocket, as well as performing basic fixes outside the station. These mechanical victories have made ready for the improvement of cutting edge frameworks that will be pivotal for future profound space missions.

Be that as it may, the excursion of the ISS has not been without its portion of misfortunes. Specialized breakdowns, like the disappointment of basic frameworks or the corruption of fundamental parts, have tried the versatility of the station and its team. One outstanding test was the unforeseen loss of tension in the Russian Soyuz space apparatus docked to the station in 2018, provoking a quick reaction from the team and ground control to recognize and redress the issue. Such occurrences highlight the significance of consistent cautiousness and the requirement for strong alternate courses of action in the unforgiving climate of room.

The political scene on Earth has additionally affected the direction of the ISS program. Moving needs, monetary limitations, and international strains have introduced difficulties to the supported responsibility of countries to the station. The victory here lies in the capacity of the worldwide accomplices to explore these difficulties and support the ISS program for north of twenty years, encouraging a persevering through tradition of collaboration in space investigation.

Looking forward, the ISS faces the test of guaranteeing a smooth progress to its next stage. Conversations about the fate of the station incorporate the chance of commercializing specific components or changing to another model of room investigation. These choices will shape the tradition of the ISS and decide how its accomplishments and illustrations learned will add to the up and coming age of human space attempts.

The victories of the ISS reach out past the limits of room investigation and science. The global coordinated effort that carried the station to completion has laid the basis for future undertakings in space. The examples gained from the ISS have educated the improvement regarding the Door, a lunar station that will act as an organizing guide for future missions toward the Moon and then some. The ISS, with its assorted group of space explorers from various countries, has turned into an image of solidarity, showing the way that humankind can accomplish striking accomplishments while cooperating towards a shared objective.

The difficulties and wins of the ISS are a microcosm of the more extensive human involvement with space investigation. They mirror our capacity to conquer difficulty, gain from our errors, and push the limits of what is conceivable.

As we commend the accomplishments of the ISS, we should likewise perceive the continuous difficulties that lie ahead. The quest for information, the headway of innovation, and the soul of investigation keep on driving humankind forward, moving people in the future to try the impossible.

All in all, the Worldwide Space Station remains as a demonstration of the unstoppable human soul and the force of global joint effort. From its commencement to its continuous tasks, the ISS has explored a perplexing trap of difficulties, arising successful through wins in innovation, science, and discretion. The tradition of the ISS stretches out a long ways past its actual presence in circle, impacting the eventual fate of room investigation and filling in as a reference point of motivation for a long time into the future. As we ponder the difficulties confronted and wins accomplished, we track down a convincing story of flexibility, resourcefulness, and the steady quest for information in the immense spread of room.

6.1 The collaborative effort in constructing and maintaining the ISS

The Worldwide Space Station (ISS) is a demonstration of the force of joint effort on a worldwide scale. From its origination to its continuous tasks, the ISS has been a striking illustration of global participation in the field of room investigation. The cooperative exertion in developing and keeping up with the ISS includes commitments from space offices and accomplices all over the planet, each bringing exceptional ability, assets, and capacities to the task.

The starting points of the ISS joint effort can be followed back to the furthest limit of the Virus War when international pressures started to defrost between the US and Russia. What was once a space race energized by political competition changed into a dream of collaboration in space. In 1998, the primary module of the ISS, the Russian Zarya module, was sent off into space. This obvious the start of an extraordinary period of joint effort among NASA and the Russian space organization, Roscosmos.

The joint effort extended past the U.S. furthermore, Russia to incorporate global accomplices like the European Space Organization (ESA), the Japan Aviation Investigation Office (JAXA), and the Canadian Space Organization (CSA). Each accomplice carried its extraordinary assets to the task, contributing both monetarily and mechanically. The

European modules, the Japanese Kibo lab, and the Canadian mechanical arm, Canadarm2, are among the striking commitments that have advanced the abilities of the ISS.

One of the early victories of the cooperative exertion was the effective gathering of the ISS in space. The measured development approach included sending off individual parts and gathering them in circle. This expected careful preparation and coordination between the worldwide accomplices. The main American module, Solidarity, was associated with Zarya in 1998, shaping the center construction of the ISS. Ensuing modules were added throughout the long term, growing the station's abilities and living space.

The joint effort confronted various specialized difficulties during the development stage. Organizing dispatches, guaranteeing similarity of various modules, and dealing with the complexities of gathering in microgravity were among the obstacles that must be survived. Be that as it may, the cooperative soul won, and the ISS step by step came to fruition as an image of global solidarity and accomplishment.

The continuous support of the ISS has been a common obligation among the global accomplices. Normal resupply missions bring fundamental arrangements, hardware, and logical instruments to the station. The strategies of these missions include careful intending to guarantee that the ISS remains completely functional and can uphold its group of space travelers and cosmonauts.

One of the critical parts of the cooperative exertion is the sharing of assets and ability. The Russian Soyuz shuttle has been a dependable workhorse for shipping space travelers and cosmonauts to and from the ISS. This coordinated effort has been pivotal, particularly during periods when the US didn't have its own maintained send off capacity. The Soyuz shuttle filled in as a life saver, guaranteeing a persistent human presence on the ISS.

The cooperative idea of the ISS is clear in its development and support as well as in the everyday exercises of the team. Space explorers and cosmonauts from various nations live and cooperate in the bound space of the ISS. This multicultural climate encourages shared grasping, regard, and brotherhood among the team individuals. It fills in as a microcosm of what can be accomplished when people from different foundations meet up with a mutual perspective.

Logical examination led on board the ISS further features the advantages of global cooperation. The station gives a novel microgravity climate that permits specialists to lead tests unrealistic on The planet. The aftereffects of these tests add to how we might interpret basic logical standards and have functional applications in fields going from medication to materials science.

The cooperative exertion stretches out to the ground control focuses that supervise the tasks of the ISS. Groups from various nations cooperate to screen the station's frameworks, plan exercises, and investigate issues. The capacity to flawlessly incorporate the endeavors of various control places mirrors the elevated degree of coordination and trust that has been laid out among the worldwide accomplices.

Regardless of the achievement and wins, the cooperative exertion in keeping up with the ISS has not been without its difficulties. International pressures on Earth have now and again stressed relations between the US and Russia, the two essential supporters of the ISS. The political scene can impact choices about the fate of the ISS program, financing assignments, and the degree of joint effort. Exploring these difficulties requires discretionary expertise and a promise to the common objectives of room investigation.

Innovative difficulties have additionally tried the cooperative exertion. The ISS works in an unforgiving climate, presented to radiation, outrageous temperatures, and the microgravity of room. Specialized breakdowns, like hardware disappointments or framework misfires, require facilitated investigating and critical thinking. The capacity of the global accomplices to cooperate consistently in addressing these difficulties addresses the strength of the cooperative structure.

As the ISS enters its third ten years of activity, conversations about its future present new difficulties and open doors for joint effort. The station has surpassed its unique arranged life expectancy, and choices about its destiny include contemplations about maintainability, cost-adequacy, and the advancing scene of room investigation. Proposition for progressing specific components of the ISS to business substances or diverting assets to new investigation drives require cautious discussion among the global accomplices.

The victories of the cooperative exertion in developing and keeping up with the ISS reach out past the domain of room investigation. The ISS fills in as an image of what can be accomplished when countries put away political contrasts in quest for shared objectives. It remains as an encouraging sign and motivation, displaying mankind's capacity to conquer difficulties through participation and creativity.

The illustrations gained from the ISS coordinated effort have suggestions for future undertakings in space investigation. As mankind focuses on aggressive objectives, for example, getting back to the Moon and in the long run sending people to Mars, the cooperative model laid out by the ISS can act as a plan. The Worldwide Space Station has given a stage to testing innovations, leading investigations, and improving the abilities expected for long-term space missions.

Looking forward, the cooperative exertion in space investigation is ready to extend much further. Global organizations are fundamental to drives, for example, NASA's Artemis program, which expects to return people to the lunar surface. The Passage, a lunar circling station, addresses one more cooperative endeavor that will include commitments from numerous countries. These drives expand upon the tradition of the ISS and show a proceeded with obligation to investigating the universe together.

All in all, the cooperative exertion in developing and keeping up with the Worldwide Space Station is a demonstration of the extraordinary force of global collaboration. From the beginning of the Virus Battle to the current period of room investigation, the ISS has resisted international limits and encouraged a feeling of solidarity among countries. The victories of collecting and working the ISS in circle have exhibited the strength of human creativity and the advantages of coordinated effort chasing after information and investigation. As the ISS keeps on orbitting the Earth, it fills in as a sign of what can be accomplished when humankind cooperates to try the impossible.

6.2 Experiences of astronauts living and working in space

The encounters of space travelers living and working in space offer an exceptional and significant point of view on the difficulties, marvels, and human versatility in the extraterrestrial climate. As mankind adventures past the bounds of Earth, the Global Space Station (ISS) has turned into a microcosm of life in space, filling in as a proving ground for the physical and mental impacts of broadened space missions.

The excursion of a space traveler starts some time before takeoff. Broad preparation, both genuinely and intellectually, is an essential for those decided to wander into space. Preparing includes figuring out how to work shuttle frameworks, leading extravehicular exercises (spacewalks), and grasping the complexities of life on board the ISS. Actual wellness is vital, as space travelers should adjust to the microgravity climate and the cost it takes on the human body.

The send off itself is a dazzling yet serious experience. The space travelers are tied into their space apparatus, feeling the vibrations and powers as the rocket drives them past Earth's air. The change from gravity to microgravity is unexpected, and the impression of weightlessness, while freeing, requires a time of change. The primary minutes in space are much of the time a combination of elation and a feeling of significant weightlessness, as space explorers float openly in the restricted quarters of their space apparatus.

Once on board the ISS, space explorers experience a novel living climate. Microgravity, or weightlessness, is one of the characterizing highlights of room life. While this shortfall of gravity considers smooth motion and a feeling of opportunity, it likewise presents difficulties to the human body. Muscles and bones go through changes in light of the decreased burden, prompting muscle decay and bone thickness misfortune. To neutralize these impacts, space travelers take part in a thorough activity routine that incorporates cardiovascular exercises and obstruction preparing.

Variation to microgravity reaches out past the actual domain. Basic exercises like eating, dozing, and individual cleanliness require changes. Food, painstakingly ready and bundled to forestall drifting endlessly, takes on another importance. Eating in microgravity includes tying down food compartments and utensils to keep away from the drifting beads of fluids. Resting quarters are individual units with hiking beds got to keep space explorers from floating capriciously during the evening.

It is a painstakingly arranged daily practice to Keep up with individual cleanliness in microgravity. Water is a valuable asset, and each drop should be utilized effectively. Space explorers utilize rinseless wipes for washing and cautiously oversee water utilization. The shortfall of a shower or running water requires creative answers for neatness and individual solace.

The mental parts of life in space are similarly critical. Space explorers live and work in a bound space, disengaged from the immeasurability of the universe. The mental difficulties of segregation, dullness, and the shortfall of normal prompts like constantly can be significant. To address these difficulties, space explorers partake in standard correspondence with friends and family and get support from a ground-based group of clinicians.

The perspective on Earth from space is a wellspring of motivation and thoughtfulness for space explorers. The outline impact, a term begat by space logician Straight to the point White, depicts the change in mindfulness and viewpoint that space explorers frequently report in the wake of seeing Earth from circle. The delicate excellence of our planet, the interconnectedness of all life, and the boundlessness of the universe pass on an enduring effect on those lucky enough to observe it.

The work on board the ISS is different and requesting. Logical examinations led in the one of a kind microgravity climate add to how we might interpret central standards in physical science, science, and materials science. Space travelers perform support undertakings, lead spacewalks to fix and overhaul frameworks, and take part in instructive effort exercises to associate with crowds on The planet.

Spacewalks, or extravehicular exercises (EVAs), are one of the most difficult and outwardly dazzling parts of a space traveler's insight. Drifting external the ISS, got by ties, space explorers lead fixes and overhauls while encompassed by the limitlessness of room. The notable picture of a space traveler outlined against the setting of Earth typifies the boldness and specialized ability expected for such undertakings.

The dependence on innovation is ubiquitous in the day to day routines of space explorers. The ISS is a mind boggling rocket with various frameworks that require checking and support. Space travelers work with trend setting innovations, from mechanical technology to logical instruments, adding to the station's usefulness and logical targets. The capacity to investigate and fix gear is an essential expertise, and space explorers go through broad preparation to deal with different possible issues.

Correspondence with Earth is worked with through an organization of satellites and ground stations. Space travelers utilize a mix of voice, video, and information transmission to remain associated with mission control and their families. The feeling of disconnection is moderated by normal correspondence, permitting space travelers to share their encounters, look for direction, and keep an association with their earthly roots.

The idea of time takes on an alternate aspect in space. The ISS circles the Earth roughly like clockwork, prompting 16 dawns and dusks every day.

The shortfall of a characteristic day-night cycle can disturb circadian rhythms, and space explorers stick to a painstakingly arranged plan that incorporates assigned work periods, exercise, recreation, and rest. Keeping a feeling of routine is fundamental for mental prosperity.

Space travelers frequently talk about the kinship and collaboration that creates among group individuals during their time on board the

ISS. The crowdedness and shared encounters make a special bond among people from various countries and foundations. The capacity to team up really is vital for the progress of the mission, and group individuals depend on each other for help in both daily practice and testing circumstances.

The re-visitation of Earth is a complex and genuinely requesting process. Once more reemergence into Earth's climate includes extraordinary intensity and tension, and space explorers experience the power of gravity. The return container, whether a Soyuz space apparatus or a business team vehicle, lands in a distant area on The planet. Recuperation groups are conveyed to help the space travelers as they straighten out to the draw of gravity and start the course of reintegration into earthly life.

The encounters of space travelers living and working in space contribute priceless experiences to how we might interpret human variation to the space climate. Logical examinations led on board the ISS have yielded disclosures with applications both in space investigation and on The planet. From propels in clinical examination to developments in materials science, the ISS fills in as a stage for pushing the limits of human information.

The individual stories of space explorers give a human aspect to space investigation. They share accounts of miracle, challenge, and self-awareness. Space travelers frequently become advocates for space investigation, stressing the significance of worldwide coordinated effort, logical revelation, and the quest for information to support mankind.

The ISS, as an image of worldwide participation, has facilitated space travelers from different countries, cultivating a feeling of solidarity among the global local area. The common encounters of the people who have lived and worked on board the ISS add to an aggregate comprehension of the conceivable outcomes and restrictions of human life past Earth.

Looking forward, as humankind focuses on additional drawn out missions to the Moon, Mars, and then some, the encounters of space travelers on the ISS act as an establishment for future investigation. The examples gained from living and working in space educate the advancement regarding innovations, preparing conventions, and mental emotionally supportive networks for the up and coming age of room voyagers.

All in all, the encounters of space explorers living and working in space address a noteworthy part in the continuous story of human investigation. From the moves of variation to microgravity to the striking perspectives on Earth from circle, space explorers give a novel point of view on

our spot in the universe. Their versatility, creativity, and devotion act as a motivation for people in the future, as mankind keeps on pushing the limits of what is conceivable in the immense span of room.

6.3 Scientific achievements and breakthroughs from ISS research

The Worldwide Space Station (ISS) remains as an image of human creativity and global coordinated effort, yet it is likewise a stage for historic logical examination. Since its origin, the ISS has filled in as an exceptional microgravity lab, giving researchers the chance to direct examinations and examinations that would be unimaginable on The planet. The logical accomplishments and forward leaps coming about because of ISS research length a great many disciplines, from key physical science to applied clinical exploration, adding to how we might interpret the universe and helping life on The planet.

One of the vital areas of logical examination on board the ISS is in the field of human wellbeing. Microgravity significantly affects the human body, and concentrating on these impacts gives bits of knowledge into physiological changes that have suggestions for the two space explorers and Terrestrial populaces. Bone thickness misfortune, muscle decay, and changes in cardiovascular capability are among the difficulties looked by space explorers living in microgravity.

To resolve these issues, researchers have led investigates the ISS to figure out the fundamental components and foster countermeasures. The utilization of cutting edge imaging methods and physiological checking has given significant information on the effect of microgravity on bone design and thickness. This examination has prompted the improvement of activity regimens and drug intercessions to alleviate bone misfortune and muscle decay during long-term space missions.

Moreover, concentrates on the ISS have added to how we might interpret cardiovascular wellbeing in space. Drawn out openness to microgravity can prompt changes in blood stream, liquid dispersion, and cardiovascular capability. By concentrating on these changes, researchers have acquired experiences into conditions, for example, orthostatic narrow mindedness, which can influence space explorers upon their re-visitation of Earth. The information acquired from ISS research has educated the advancement regarding methodologies to keep up with cardiovascular wellbeing during space missions and has significance for resolving comparable issues in patients on The planet.

The impacts of room travel on the human insusceptible framework have likewise been a subject of extraordinary concentrate on the ISS. Living in a restricted space with a shut ecological framework presents difficulties to the resistant framework, and understanding what microgravity means for safe capability is urgent for the wellbeing and prosperity of space travelers. ISS tests have revealed insight into changes in safe cell capability, cytokine creation, and the vulnerability to diseases in space.

Past human wellbeing, the ISS fills in as a stage for state of the art research in key physical science and materials science. Microgravity permits researchers to concentrate on peculiarities that are veiled by Earth's gravity, prompting a more profound comprehension of the basic laws of material science. Probes the ISS have investigated liquid elements, ignition processes, and the way of behaving of materials in space.

In the domain of liquid elements, the ISS has given a special climate to concentrate on the way of behaving of liquids in microgravity. Fine stream tests, for instance, have analyzed how fluids move without gravity, with suggestions for shuttle liquid administration and life emotionally supportive networks. Such examination has pragmatic applications in planning more effective liquid frameworks for future space missions.

Ignition probes the ISS have progressed how we might interpret fire conduct in microgravity. The shortfall of lightness driven convection permits researchers to seclude and concentrate on the essential cycles of burning. This examination has suggestions for space apparatus well-being and the improvement of more proficient burning advancements on The planet.

Materials science probes the ISS have investigated the properties of materials in microgravity, uncovering experiences that can prompt better materials for different applications. From semiconductor gem development to the investigation of liquid conduct in permeable media, ISS research has added to the improvement of new materials with up-graded properties. These materials might track down applications in gadgets, fabricating, and different ventures on The planet.

The exceptional microgravity climate of the ISS has additionally demonstrated important for research in central material science, includ-ing tests that test the constraints of how we might interpret the universe. Accuracy tests in regions, for example, antimatter and basic powers have been led on the ISS. These tests add to our investigation of the principal building blocks of issue and the powers that administer the universe.

Besides, the ISS has been a stage for space-based galactic perceptions. Instruments on the station, away from the World's air, give a more clear perspective on far off divine items. The Alpha Attractive Spectrometer (AMS-02), for instance, is a molecule material science try mounted on the outside of the ISS. It is intended to concentrate on grandiose beams, looking for antimatter and dull matter particles, and contribut-ing important information to how we might interpret the universe's arrangement.

The headways in life sciences and innovation coming about because of ISS research have expansive ramifications for future space investi-gation tries. As humankind focuses on additional drawn out missions, including the Moon, Mars, and then some, the illustrations gained from the ISS are pivotal for guaranteeing the wellbeing and prosperity of space travelers and the progress of investigation missions.

One more area of huge logical accomplishment from ISS research is in the domain of Earth and ecological sciences. The one of a kind vantage reason behind the ISS, circling the Earth roughly like clockwork, gives a chance to notice our planet in manners that were beforehand unthinkable. Earth perception instruments on the ISS catch information on environment, weather conditions, and cataclysmic events, adding to a superior comprehension of Earth's dynamic frameworks.

The Hyperspectral Imager for the Waterfront Sea (HICO) is an illustration of an Earth perception instrument on the ISS. HICO catches high-goal pictures of seaside locales, assisting researchers with observing changes in water quality, sea flows, and the wellbeing of marine biological systems. The information gathered by HICO and comparative instruments add to natural checking, asset the board, and debacle reaction endeavors.

The Cloud-Spray Transport Framework (Felines) is one more Earth perception instrument on the ISS intended to concentrate on the appropriation of vapor sprayers in Earth's climate. By understanding spray fixations and developments, researchers can acquire experiences into air quality, environmental change, and the effect of human exercises on the climate.

ISS research plays likewise had an impact in concentrating on the World's environment and the impacts of environmental change. Instruments, for example, the Stratospheric Spray and Gas Trial (SAGE) III have been sent on the ISS to screen the ozone layer, ozone depleting substance focuses, and other barometrical boundaries. The information gathered add to environment models and illuminate worldwide endeavors to address natural difficulties.

The interdisciplinary idea of ISS research is exemplified by the many analyses that overcome any barrier between life sciences and Studies of the planet. For instance, examinations concerning plant science on the ISS have given experiences into how plants answer microgravity and how they can be utilized in life emotionally supportive networks for future long-term space missions. Simultaneously, these investigations have useful applications on The planet, offering possible answers for practical farming and ecological protection.

The coordinated effort between global accomplices on the ISS has additionally stretched out to Earth and ecological sciences. The information gathered from Earth perception instruments benefit researchers and policymakers around the world, encouraging a worldwide comprehension of our planet's interconnected frameworks. The ISS fills in as a stage for helpful endeavors to address ecological difficulties that rise above public limits.

In the area of innovation improvement, the ISS has been a testbed for new advances and designing arrangements with applications both in space and on The planet. The limitations of the space climate request developments in materials, mechanical technology, and life emotionally supportive networks, driving headways that have viable ramifications for different enterprises.

One outstanding model is the Automated Refueling Mission (RRM) on the ISS. RRM is a progression of examinations that exhibit satellite overhauling capacities utilizing mechanical innovations. The capacity to refuel and fix satellites in circle has huge ramifications for expanding the life expectancy of satellites and lessening space flotsam and jetsam. The information acquired from RRM adds to the advancement of independent mechanical frameworks with applications in satellite adjusting, space investigation, and Earth-based ventures.

Headways in 3D printing innovation have likewise been tried on the ISS. The Added substance Assembling Office (AMF) permits space explorers to deliver devices and parts on-request, diminishing the requirement for broad extra parts stock. The effective utilization of 3D imprinting in space has suggestions for future long-span missions where asset productivity is principal.

Moreover, research on the ISS has added to the improvement of regenerative life emotionally supportive networks. Shut circle frameworks that reuse air and water are fundamental for supporting human existence during expanded space missions. The Water Recuperation Framework on the ISS, for instance, sanitizes wastewater, making it appropriate for utilization. This innovation has applications in water-scant areas on The planet, offering answers for maintainable water the executives.

In the domain of telemedicine, ISS research has educated the improvement regarding far off clinical checking advances. The difficulties of giving medical care in the restricted and disengaged climate of room equal those looked in remote or out of reach locales on The planet. Telemedicine arrangements tried on the ISS add to further developing medical services conveyance in underserved regions and during crises.

The logical accomplishments and forward leaps from ISS research reach out past the bounds of the space station, impacting different fields and helping humankind overall. The coordinated effort among worldwide accomplices, researchers, specialists, and space explorers has made a tradition of development and disclosure. As the ISS proceeds with its central goal in circle, it stays a signal of human accomplishment and a demonstration of the extraordinary force of logical investigation.

Looking forward, the illustrations gained from ISS examination will shape the fate of room investigation and advise the plan regarding environments for future missions to the Moon, Mars, and then some. The interdisciplinary idea of ISS research, crossing human wellbeing, crucial physical science, Studies of the planet, and innovation advancement, highlights the comprehensive methodology expected for tending to the difficulties of room investigation and propelling comprehension we might interpret the universe. The tradition of logical accomplishments from the ISS fills in as a motivation for the up and coming age of researchers, specialists, and pilgrims, rousing them to push the limits of information and proceed with mankind's excursion into the universe.

The Global Space Station (ISS) remains as a circling lab, cultivating cooperation among countries and driving logical forward leaps that have sweeping ramifications for both space investigation and life on The planet. The extraordinary microgravity climate of the ISS has furnished researchers with a phenomenal stage to lead trials and examinations across a range of disciplines. From human wellbeing to principal physical science, Earth and natural sciences, and mechanical developments, the leap forwards from ISS research address a demonstration of the force of worldwide participation and human inventiveness.

A critical domain of leap forwards emerging from ISS research lies in the field of human wellbeing. Microgravity's impacts on the human body, for example, bone thickness misfortune, muscle decay, and adjustments in cardiovascular capability, present difficulties for space explorers setting out on lengthy term space missions. In any case, the ISS has turned into a demonstrating ground for creating countermeasures and acquiring experiences into relieving these medical problems.

Specialists have utilized the ISS to research bone wellbeing in microgravity. Tests including progressed imaging strategies and physiological observing have given important information on bone construction and thickness changes in space. These discoveries have prompted the improvement of designated practice regimens and drug mediations to balance bone misfortune, guaranteeing the prosperity of space travelers during broadened space missions.

Muscle decay, one more outcome of delayed openness to microgravity, has been a focal point of ISS research. Space explorers take part in thorough work-out schedules to keep up with bulk and capability. The information acquired from these examinations has added to the strength of space travelers as well as educated earthbound restoration rehearses for patients on Earth confronting muscle-related difficulties.

Cardiovascular wellbeing is a basic part of human prosperity, and the ISS has been instrumental in propelling comprehension we might interpret its elements in space. Delayed spaceflight can prompt changes in blood stream, liquid conveyance, and cardiovascular capability. ISS research has given bits of knowledge into conditions like orthostatic narrow mindedness, a peculiarity influencing space travelers upon their re-visitation of Earth. These experiences add to systems for keeping up with cardiovascular wellbeing during space missions and proposition likely applications for resolving comparable issues in patients on The planet.

The effect of microgravity on the human safe framework has likewise been a subject of broad concentrate on board the ISS. The shut climate of a rocket difficulties the resistant framework, and ISS tests have given significant information on changes in safe cell capability, cytokine creation, and helplessness to diseases in space. These discoveries have suggestions for the two space explorers and clinical science on The planet, adding to how we might interpret resistant reactions in restricted or confined conditions.

In the domain of innovation improvement, the ISS has filled in as a testbed for imaginative arrangements with applications both in space and on The planet. The Automated Refueling Mission (RRM) is an illustration of mechanical leap forwards on the ISS. RRM exhibits satellite adjusting capacities utilizing automated advancements, preparing for broadening the life expectancy of satellites through refueling and fix in circle. The outcome of RRM has suggestions for lessening space flotsam and jetsam and working on satellite manageability.

Headways in 3D printing innovation have likewise tracked down a home on the ISS. The Added substance Assembling Office (AMF) permits space travelers to deliver instruments and parts on-request, decreasing the requirement for broad extra parts stock. The effective utilization of 3D imprinting in space has viable ramifications for future space missions where asset productivity is vital, as well as applications in enterprises on The planet.

The ISS has been a point of convergence for space-based galactic perceptions, adding to leap forwards in how we might interpret the universe. Instruments like the Alpha Attractive Spectrometer (AMS-02) are mounted on the outside of the ISS to concentrate on infinite beams, look for antimatter, and explore dim matter particles. These perceptions give basic information to propelling our insight into the key structure blocks of the universe.

In the field of Earth and ecological sciences, the ISS has turned into an important stage for concentrating on our planet from space. Earth perception instruments catch high-goal pictures and information on environment, atmospheric conditions, and cataclysmic events. The Hyperspectral Imager for the Seaside Sea (HICO) is an illustration of an Earth perception instrument that screens beach front areas, offering bits of knowledge into water quality, sea flows, and marine biological system wellbeing. The information gathered from such instruments add to natural checking, asset the board, and calamity reaction endeavors.

ISS research has likewise assumed a vital part in concentrating on the World's environment and the impacts of environmental change. Instruments like the Stratospheric Spray and Gas Examination (SAGE) III screen the ozone layer, ozone depleting substance fixations, and environmental boundaries. The information add to environment models and backing worldwide endeavors to address ecological difficulties. The interdisciplinary idea of ISS research is obvious in the joint efforts between life sciences and Studies of the planet, where concentrates on in plant science offer experiences into maintainable horticulture and natural preservation.

The leap forwards from ISS research are not bound to the domain of science and innovation; they stretch out to the human experience and our comprehension of life in space. Space explorers on board the ISS have given firsthand records of the mental and profound parts of living and working in space. The outline impact, a significant change in mindfulness revealed by space explorers after seeing Earth from space, highlights the extraordinary idea of the space insight.

From a human wellbeing viewpoint, the mental difficulties of segregation, control, and the shortfall of normal day-night cycles have been concentrated on the ISS. Telemedicine arrangements tried in space add to further developing medical care conveyance in remote or out of reach districts on The planet. The encounters of space explorers living and dealing with the ISS act as an establishment for grasping the effect of broadened space missions on emotional wellness and prosperity.

The tradition of ISS research stretches out past its flow functional stage. As humankind plans ahead for space investigation, including missions to the Moon, Mars, and then some, the leap forwards from ISS exploration will educate the plan regarding territories, advances, and countermeasures fundamental for the wellbeing and security of space travelers. The illustrations gained from the ISS add to the diagram for supported human presence in space, directing the up and coming age of room investigation tries.

Chapter 7

Beyond Earth's Orbit: Interplanetary Exploration

The mission for investigation has for some time been an inherent part of human instinct. From the earliest long stretches of exploring the immense seas to the contemporary quest for the universe, the human soul has longed to push the limits of the known. Past Earth's circle, interplanetary investigation coaxes as the following boondocks, an excursion that rises above the bounds of our home planet and moves us into the strange domains of the planetary group.

Interplanetary investigation isn't just a logical undertaking; it is a demonstration of mankind's voracious interest and assurance. The appeal of far off planets, moons, and divine bodies has dazzled the creative mind of ages, cultivating a dauntless soul of experience. The inestimable expressive dance of our planetary group unfurls before us, offering a material whereupon we paint the fantasies of tomorrow. To leave on an excursion past Earth's circle is to embrace the obscure, to challenge the restrictions of our mechanical ability and scholarly sharpness.

The excursion to the external spans of the nearby planet group starts with a key inquiry: why investigate? The response lies in the quintessence of human interest, the tenacious quest for information, and the craving to figure out our position in the universe. Earth, our support, has fed the seeds of progress, however the endlessness of room allures us to broaden our scope. Interplanetary investigation is driven by the journey for disclosure, the quest for extraterrestrial life, and the yearning to unwind the secrets of the universe.

Mars, the red planet, remains as a tempting objective for interplanetary investigation. The charm of Mars lies in its nearness to Earth as well as in its true capacity as a harbinger of life past our home planet. The Martian scene, with its corroded tones and confounding elements, has been a subject of interest for stargazers and researchers the same. The fantasy about sending people to Mars has been a tenacious subject in the domain of room investigation, and endeavors are in progress to change this fantasy into a substantial reality.

The innovative difficulties of interplanetary travel are impressive. Defeating the tremendous distances between planets requires advancements in drive frameworks, route, and life support innovations. The improvement of cutting edge shuttle equipped for supporting human existence for expanded periods is a basic achievement in the excursion past Earth's circle. Atomic impetus, particle drives, and other state of the art innovations are being investigated to push space apparatus through the astronomical spread, beating the impediments of ordinary rocket impetus.

As we put our focus on interplanetary objections, the idea of maintainability takes on another aspect. The difficulties of asset use, reusing, and life support become foremost in the cruel climate of room. Making shut circle environments that copy Earth's self-supporting cycles is fundamental for long-length missions past our planet. Interplanetary natural surroundings should be outfitted with the resources to produce food, reuse squander, and give a reasonable climate to the bold wayfarers wandering into the infinite unexplored world.

The human component in interplanetary investigation adds a layer of intricacy and versatility to the mission. The mental and physiological impacts of long haul space travel present huge difficulties that require exhaustive arrangements. The disengagement, repression, and microgravity conditions experienced by space explorers request developments in clinical science, psychological wellness support, and actual wellness regimens. Understanding and moderating the effects of room travel on the human body are pivotal for the outcome of interplanetary missions.

The Worldwide Space Station (ISS) fills in as a microcosm of the difficulties and arrangements related with long-term space missions. The examples gained from the ISS, with its global team and cooperative endeavors, give significant bits of knowledge into the complexities of living and working in space. The change from the controlled climate of low Earth circle to the erratic states of interplanetary space requires a quantum jump in how we might interpret space home.

Past Mars, the gas monsters Jupiter and Saturn allure as divine goliaths with their company of moons. The frigid moons of Europa, Ganymede, Enceladus, and Titan hold the commitment of opening the mysteries of subsurface seas and the potential for extraterrestrial life. Automated missions, like the Europa Trimmer and the Dragonfly mission to Titan, are ready to investigate these far off universes, preparing for future human investigation.

Jupiter's moon Europa, with its smooth frigid surface covering a subsurface sea, has caught the creative mind of researchers and space devotees. The possibility of finding indications of something going on under the surface underneath the ice has impelled the Europa Trimmer mission, a cooperation among NASA and the European Space Organization (ESA). The space apparatus will lead point by point surveillance of Europa's surface, concentrating on creation and looking for signs might allude to the presence of life in the subsurface sea.

Saturn's biggest moon, Titan, flaunts a thick environment and a scene specked with lakes and waterways of fluid methane and ethane. The Dragonfly mission, a rotorcraft intended to investigate Titan's different territories, plans to study prebiotic science and survey the moon's livability. Dragonfly's capacity to fly starting with one area then onto the next permits it to cross immense distances on Titan's surface, giving a one of a kind viewpoint on this puzzling moon.

The external planets, Uranus and Neptune, present an exceptional arrangement of difficulties and potential open doors for investigation. Their far off circles and the shortage of daylight in the external nearby planet group require imaginative ways to deal with power age and rocket plan. Automated missions to these ice monsters could open the insider facts of their airs, attractive fields, and ring frameworks, giving experiences into the arrangement and development of the nearby planet group.

The possibility of mining the bountiful assets of space rocks and comets adds a financial aspect to interplanetary investigation. These divine bodies, leftovers of the early nearby planet group, harbor important metals, water, and different assets that could fuel future space attempts. Space rock mining missions, for example, NASA's OSIRIS-REx and Japan's Hayabusa2, mean to gather tests from space rocks and return them to Earth for examination. The information acquired from these missions could make ready for supported human presence past Earth's circle.

The excursion past Earth's circle isn't exclusively the area of government space organizations. Privately owned businesses, energized by pioneering vision and mechanical advancement, are assuming an undeniably unmistakable part in molding the eventual fate of room investigation. Organizations like SpaceX, Blue Beginning, and others are creating reusable rocket innovation, business space environments, and aggressive designs for lunar and interplanetary missions. The rise of a lively space economy holds the possibility to democratize admittance to space and introduce another period of investigation and revelation.

The Moon, Earth's divine buddy, possesses a focal job in the guide to interplanetary investigation. Lunar missions act as venturing stones, giving a proving ground to innovations and methodologies fundamental for wandering farther into the nearby planet group. NASA's Artemis program expects to return people to the Moon, laying out a reasonable presence and planning for future missions to Mars and then some. The lunar passage, a space station in circle around the Moon, will act as an organizing point for interplanetary missions, working with the get together of rocket and empowering group moves.

The use of lunar assets further improves the plausibility of interplanetary investigation. The Moon's regolith, or lunar soil, contains significant minerals and water ice that could be extricated to help future missions. In-situ asset usage (ISRU) innovations, like lunar mining and water extraction, are basic to the vision of making a self-supporting framework in space. The Moon, with its nearness and openness, turns into a pivotal proving ground for refining these innovations prior to setting out on more aggressive interplanetary excursions.

As mankind expands its venture into the universe, the lawful and moral contemplations of room investigation come to the very front. The Space Arrangement, endorsed in 1967, structures the groundwork of worldwide space regulation, accentuating the serene utilization of space and disallowing the appointment of divine bodies. As business elements enter the space field, inquiries of asset proprietorship, ecological effect, and the guideline of room exercises become progressively complicated. A cooperative and comprehensive way to deal with space administration is fundamental to guarantee the mindful and economical investigation of the planetary group.

The job of man-made consciousness (computer based intelligence) in interplanetary investigation couldn't possibly be more significant. From independent shuttle route to information investigation and mission arranging, artificial intelligence advances improve the productivity and capacities of room missions. AI calculations process tremendous measures of information, empowering continuous independent direction and versatile reactions to unanticipated difficulties. As we adventure past Earth's circle, artificial intelligence turns into a basic accomplice in our journey to investigate the universe.

The progression of interplanetary investigation requires worldwide joint effort on a remarkable scale. The Artemis Accords, a bunch of standards for lunar investigation, represent the soul of collaboration among spacefaring countries. Shared foundation, cooperative exploration, and the pooling of assets prepare for a future where humankind all in all investigates the wildernesses of room. The Global Space Station fills in as a model for worldwide participation, showing the potential for different countries to meet up chasing shared objectives.

Interplanetary investigation is intrinsically attached to the more extensive setting of room science and astronomy. The investigation of exoplanets, divine bodies past our nearby planet group, gives significant experiences into the variety of planetary frameworks and the

circumstances helpful for life. The James Webb Space Telescope, ready to send off before long, will upset how we might interpret the universe by looking into the climates of exoplanets and unwinding the secrets of their structure.

The quest for extraterrestrial life stays a main thrust behind interplanetary investigation. The revelation of microbial life on Mars or the subsurface expanses of cold moons would have significant ramifications for how we might interpret the beginnings of life in the universe. The investigation of outrageous conditions on The planet, from remote ocean aqueous vents to acidic lakes, fills in as analogs for expected living spaces on different planets. As we adventure into the universe, the journey for life past Earth remains as a binding together subject that rises above public limits and logical disciplines.

The effect of interplanetary investigation stretches out past the domains of science and innovation. The motivation got from arriving at new wildernesses touches off the human creative mind and fills the quest for information. Instructive effort and public commitment assume a pivotal part in cultivating a worldwide appreciation for the marvels of room investigation. The pictures and information sent from far off planets and moons enamor the public's creative mind, motivating the up and coming age of researchers, designers, and wayfarers.

The excursion past Earth's circle is a demonstration of human strength, resourcefulness, and the unfathomable soul of investigation.

An excursion rises above the restrictions of our natural presence, welcoming us to ponder our position in the universe. As we put our focus on far off planets, moons, and space rocks, we set out on a journey that goes past the actual bounds of our space apparatus. It is an excursion of the psyche, a journey to unwind the secrets of the universe and to graph a course toward a future where the limits of investigation are restricted exclusively by the scopes of our aggregate creative mind.

7.1 Exploration of other celestial bodies using rockets

The investigation of other divine bodies utilizing rockets addresses a zenith of human accomplishment, a demonstration of our persistent interest and want to unwind the secrets of the universe. Rockets, with their capacity to break liberated from Earth's gravitational hug, have turned into the way to arriving at far off planets, moons, and then some. This excursion into the obscure has extended how we might interpret the universe as well as pushed the limits of innovation, designing, and human perseverance.

The coming of rocketry for the purpose of room investigation can be followed back to the mid-twentieth 100 years, a period set apart by extraordinary international contentions and the competition to overcome the last boondocks. The summit of these endeavors was the send off of the primary counterfeit satellite, Sputnik 1, by the Soviet Association in 1957. This notable occasion denoted the beginning of the space age and set up for another period of investigation.

Following Sputnik, both the US and the Soviet Association escalated their endeavors to push the limits of room investigation. The Virus War contention powered a progression of notable accomplishments, remembering the principal human for space, Yuri Gagarin, and the famous Apollo moon missions drove by NASA. The Apollo program, with its titanic Saturn V rockets, showed the capacity of people to go past Earth's circle and gone to another heavenly body.

The development of rocket innovation has been a ceaseless course of development and refinement. From the beginning of fluid powered rockets spearheaded by visionaries like Robert H. Goddard to the strong and complex send off vehicles of today, the direction of rocket improvement has been pushed by logical creativity and mechanical progressions. The change from substance rockets to more proficient and supportable impetus frameworks, for example, particle drives and atomic drive, addresses the continuous mission to upgrade the abilities of room investigation.

Mars, frequently alluded to as the "Red Planet," has been a point of convergence of interplanetary investigation utilizing rockets. The excursion to Mars includes complex orbital elements, exact route, and high level life emotionally supportive networks to support space explorers on the long-span journey. The send off windows for Mars missions happen around at regular intervals when the overall places of Earth and Mars consider the most proficient exchange of shuttle. This synodic period makes way for aggressive missions to concentrate on the Martian surface, environment, and expected indications of past or present life.

NASA's Mars wanderers, including Sojourner, Soul, Opportunity, Interest, and Constancy, address a heredity of mechanical pioneers that have crossed the Martian scene, directing examinations, investigating soil tests, and catching stunning pictures of the Martian landscape. These wanderers, sent off on strong rockets, have given priceless information that has developed how we might interpret Mars and educated the plan regarding future human missions.

The journey to send people to Mars is a great endeavor that requires another age of rockets, space apparatus, and life emotionally supportive networks. The Space Send off Framework (SLS), created by NASA, remains as a demonstration of the scale and desire of interplanetary investigation. With its strong motors and transcending structure, the SLS is intended to convey space travelers past Earth's circle, making ready for manned missions to the Moon, Mars, and then some.

The idea of room colonization and the foundation of human living spaces on other heavenly bodies have gotten forward movement lately. Making manageable stations on the Moon or Mars includes not just conquering the specialized difficulties of rocket travel yet additionally tending to the intricacies of life support, asset use, and the mental prosperity of occupants. Rockets, as the essential method for transportation, become the helps interfacing Earth to these extraterrestrial settlements.

The Moon, Earth's nearest heavenly neighbor, has for quite some time been an objective for investigation and fills in as a proving ground for innovations that will empower future interplanetary missions. The Artemis program, drove by NASA, plans to return people to the lunar surface and lay out a feasible presence. The improvement of the Space Send off Framework and the Orion space apparatus, sent off on strong rockets, is a critical stage in understanding the vision of a human revisitation of the Moon.

The usage of rockets for lunar investigation reaches out past government space organizations. Privately owned businesses, like SpaceX and Blue Beginning, are effectively chasing after lunar desires. SpaceX's Starship, a completely reusable shuttle and rocket framework, imagines lunar missions as well as the vehicle of people to Mars and different objections in the nearby planet group. Blue Beginning's Blue Moon lunar lander addresses another private-area attempt to open the capability of the Moon for logical exploration and future human residence.

The job of global coordinated effort in lunar investigation couldn't possibly be more significant. The Lunar Door, a space station in circle around the Moon, fills in as a cooperative exertion including NASA, the European Space Office (ESA), and other worldwide accomplices. This door, sent off into space on rockets, will work with team moves, act as an organizing point for lunar missions, and add to the more extensive objectives of interplanetary investigation.

Space rocks, rough remainders of the early planetary group, present fascinating open doors for investigation and asset usage. Rockets outfitted with modern instrumentation and automated frameworks are entrusted with arriving at these heavenly bodies to unwind their secrets and open their true capacity. Missions, for example, NASA's OSIRIS-REx and Japan's Hayabusa2 have effectively rendezvoused with space rocks, gathered examples, and are coming back to Earth, conveyed by the force of rockets.

The use of rockets for space rock mining addresses a boondocks that holds the commitment of immense assets for future space tries. Water-rich space rocks, specifically, are of interest because of the possibility to remove water for life emotionally supportive networks and the creation of rocket fuel. The monetary practicality of space rock mining, made conceivable by cutting edge rocket innovation, could alter the space business and open new roads for supported human presence past Earth.

The external planets of our planetary group, with their different moons and extraordinary elements, allure as focuses for investigation utilizing rockets. Jupiter, the biggest planet, has an entourage of intriguing moons, including Europa, Ganymede, and Io. The Galileo rocket, sent off on board the Space Transport Atlantis, gave significant bits of knowledge into Jupiter and its moons. The investigation of these far off universes includes multifaceted orbital moves and exact computations to use gravity helps from various planetary bodies.

Europa, one of Jupiter's moons, has collected specific consideration because of the potential for a subsurface sea underneath its frigid outside layer. Rockets conveying shuttle outfitted with cutting edge logical instruments, for example, the Europa Trimmer mission, intend to concentrate on the moon's surface and accumulate information that might illuminate future missions to investigate its subsurface sea. The test of arriving at far off moons like Europa highlights the significance of strong rockets fit for executing complex directions.

Saturn, decorated with its grand ring framework, has been the focal point of investigation utilizing rockets. The Cassini-Huygens mission, a coordinated effort between NASA, ESA, and the Italian Space Organization, gave extraordinary perspectives on Saturn, its rings, and its different moons. The Huygens test, conveyed by the Cassini shuttle, dropped to the outer layer of Titan, Saturn's biggest moon, uncovering a world with lakes and waterways of fluid methane and ethane.

The Dragonfly mission, set to investigate Titan, embodies the utilization of rockets to arrive at far off objections. Dragonfly, a rotorcraft intended to fly in Titan's thick environment, will examine the moon's different territory and study its prebiotic science. The test of sending off, exploring, and arriving on a far off moon like Titan requires refined rocket innovation and exact mission arranging.

Uranus and Neptune, the ice monsters of the external planetary group, present remarkable difficulties and valuable open doors for investigation. Rockets, outfitted with cutting edge instrumentation, are expected to arrive at these far off planets and study their climates, attractive fields, and ring frameworks. Automated missions to Uranus and Neptune could divulge mysteries about the arrangement and development of our planetary group, adding to our more extensive comprehension of planetary frameworks in the world.

The mix of man-made reasoning (artificial intelligence) into rocket innovation has become progressively predominant in space investigation. Computer based intelligence calculations are utilized in mission arranging, independent route, and information examination, upgrading the proficiency and capacities of room missions. AI models process huge measures of information, empowering constant navigation and versatile reactions to unanticipated difficulties. As we broaden our venture into the universe, the organization among rockets and man-made intelligence turns into a foundation of interplanetary investigation.

The lawful and moral contemplations of rocket-based investigation stretch out past the specialized and logical domains. The Space Arrangement, approved by various countries, lays out the standards of serene use and non-allotment of space. As business elements and confidential endeavors become more engaged with space exercises, inquiries of asset possession, natural effect, and space administration become the dominant focal point. The requirement for a cooperative and comprehensive way to deal with space regulation becomes fundamental as mankind adventures farther into the nearby planet group.

7.2 Robotic missions to Mars, Venus, and beyond

Mechanical missions to investigate the planets of our planetary group have been at the very front of logical revelation, pushing the limits of our insight and divulging the secrets of these far off universes. Among these heavenly bodies, Mars and Venus definitely stand out of researchers and space organizations because of their nearness to Earth and the potential for experiences into the circumstances fundamental forever. Past Mars and Venus, the investigation of gas goliaths like Jupiter and Saturn, as well as frosty moons and far off space rocks, has extended how we might interpret the different scenes and conditions that exist inside our planetary group.

Mars, frequently alluded to as the "Red Planet," has been an essential focal point of mechanical investigation. The charm of Mars lies in its true capacity as a host for past or present life, as well as its similitudes to Earth regarding geography and environment. Mechanical missions to Mars have been instrumental in concentrating on its surface, climate, and land highlights.

NASA's Mars meanderers, including Soul, Opportunity, Interest, and Determination, address a genealogy of mechanical voyagers that have essentially progressed how we might interpret the Martian climate. These meanderers, sent off on board strong rockets, have navigated the Martian surface, led trials, and sent back significant information and pictures.

The Interest meanderer, for example, arrived on Mars in 2012 and has been investigating the Hurricane Hole, concentrating in the world's topography and looking for indications of past livability. Its high level logical instruments, including a stone disintegrating laser and a drill for gathering tests, grandstand the capacities of mechanical missions to direct definite investigations of Martian landscape.

Persistence, the furthest down the line expansion to the Martian wanderer family, landed in the world's surface in 2021. Furnished with significantly more modern instruments, including an example storing framework, Persistence means to look for indications of old life and gather rock and soil tests that might be gotten back to Earth by future missions.

Notwithstanding meanderers, orbiters play had a significant impact in Martian investigation. These shuttle, sent off into space around Mars on rockets, give an elevated perspective of the planet and direct remote detecting to dissect its surface, air, and environment. Orbiters like the Mars Surveillance Orbiter (MRO) and the Mars Climate and Unstable Development (Expert) mission have fundamentally extended how we might interpret Mars.

The investigation of Mars stretches out past its surface and environment to the quest for subsurface water and the possible livability of the planet. The Understanding lander, which arrived on Mars in 2018, conveys instruments intended to test the Martian inside, including a seismometer to recognize marsquakes and an intensity stream test to gauge the planet's interior temperature.

Venus, Earth's adjoining planet, has likewise been an objective for mechanical investigation. In spite of its cold circumstances, including a thick air of sulfuric corrosive and surface temperatures sufficiently hot to liquefy lead, Venus presents one of a kind open doors for logical examination. Early missions, for example, NASA's Trailblazer Venus and the Soviet Association's Venera program, gave introductory experiences into Venusian geography and climate.

All the more as of late, NASA's Parker Sun powered Test, while essentially intended to concentrate on the Sun, performed gravity helps during its flybys of Venus, adding to how we might interpret the planet's climate and attractive field. Future missions, like NASA's VERITAS (Venus Emissivity, Radio Science, InSAR, Geography, and Spectroscopy) and ESA's Imagine, plan to additionally investigate Venus and unwind the secrets of its geography and development.

Past the rough planets of the inward planetary group, gas goliaths like Jupiter and Saturn coax for investigation. The Juno mission, sent off by NASA and conveyed into space by a strong rocket, has been circling Jupiter beginning around 2016. Juno's set-up of instruments is intended to concentrate in the world's organization, gravity field, attractive field, and polar magnetosphere. The information gathered by Juno adds to how we might interpret Jupiter's development and the cycles that administer its dynamic environment.

Saturn, with its staggering ring framework, has been the focal point of the Cassini-Huygens mission. Sent off on a rocket, the Cassini shuttle entered circle around Saturn in 2004 and gave exceptional perspectives in the world, its rings, and its different moons. The Huygens test, conveyed by Cassini, slid to the outer layer of Titan, Saturn's biggest moon, uncovering a scene with lakes and streams of fluid methane and ethane.

The investigation of cold moons inside our nearby planet group addresses one more element of mechanical missions. Europa, one of Jupiter's moons, is quite compelling because of the expected presence of a subsurface sea underneath its frosty hull. The Europa Trimmer mission, set to send off on a strong rocket, plans to concentrate on Europa's surface and subsurface to survey its expected livability.

Enceladus, a moon of Saturn, has likewise enraptured researchers with its springs ejecting from underneath its frigid surface. The Cassini space apparatus identified these crest during its central goal, raising the chance of a subsurface sea on Enceladus. Future missions might target Enceladus to additionally research its true capacity for facilitating microbial life.

Space rocks, little rough bodies that circle the Sun, have been the subject of mechanical missions pointed toward understanding the early planetary group and the structure blocks of planets. NASA's OSIRIS-REx and Japan's Hayabusa2 missions both include space apparatus sent off on board rockets to meet with and gather tests from space rocks.

OSIRIS-REx, sent off in 2016 on a Chart book V rocket, showed up at the close Earth space rock Bennu in 2018. The space apparatus gathered examples from the space rock's surface and is booked to return them to Earth before very long. Likewise, Hayabusa2, sent off on a Japanese H-IIA rocket, gathered examples from the space rock Ryugu and is returning to Earth, conveying important material that might give experiences into the early nearby planet group.

The investigation of the external spans of our nearby planet group has been worked with by mechanical missions focusing on the huge space past the circle of Neptune. The Explorer tests, sent off on strong Titan III and Titan IIIE rockets, set out on a fantastic visit through the external planets in the last part of the 1970s and mid 1980s. Explorer 1 and Explorer 2 gave uncommon perspectives on Jupiter, Saturn, Uranus, and Neptune, and both shuttle proceed with their excursions into interstellar space.

The New Skylines mission, sent off on a Chart book V rocket in 2006, led a noteworthy flyby of Pluto in 2015, giving the principal close-up pictures of this far off and cryptic world. Past Pluto, New Skylines proceeds with its direction toward the external ranges of the Kuiper Belt, concentrating on other little, frosty bodies in the external nearby planet group.

The use of man-made consciousness (artificial intelligence) has become progressively common in mechanical space missions. Man-made intelligence calculations are utilized in mission arranging, independent route, and information examination, improving the effectiveness and abilities of these missions. AI models process immense measures of information, empowering ongoing independent direction and versatile reactions to unexpected difficulties. As we broaden our venture into the universe, the mix of simulated intelligence into mechanical missions turns into a foundation of room investigation.

The legitimate and moral contemplations of mechanical space investigation stretch out past the specialized and logical domains. The Space Settlement, approved by various countries, lays out the standards of quiet use and non-assignment of space. As automated missions become more refined and investigate different divine bodies, inquiries of asset possession, natural effect, and space administration become the overwhelming focus. The requirement for a cooperative and comprehensive way to deal with space regulation becomes fundamental as mankind adventures farther into the nearby planet group.

Public commitment and instructive effort assume a critical part in the progress of mechanical missions. The pictures caught by shuttle, the logical disclosures made in the universe, and the pivotal snapshots of investigation spellbind the creative mind of individuals all over the planet. The capacity to share the marvels of room investigation, worked with by mechanical missions sent off on strong rockets, cultivates a worldwide appreciation for the significance of logical revelation and moves the up and coming age of researchers, designers, and voyagers.

7.3 The significance of interplanetary exploration in understanding the cosmos

Interplanetary investigation, the undertaking to test the secrets of our adjoining planets and heavenly bodies past Earth's circle, holds significant importance in our journey to grasp the universe. Past the logical and mechanical progressions it encourages, interplanetary investigation addresses the actual embodiment of human interest, the voracious drive to unwind the mysteries of the universe and understand our place inside it. This investigation grows our logical information as well as sparkles philosophical consideration about our reality, the beginnings of life, and the endlessness of the infinite embroidery.

One of the essential inspirations for interplanetary investigation lies chasing replies to basic inquiries regarding the beginnings and advancement of our planetary group. Every planet, moon, and space rock fills in as a period case, safeguarding signs about the circumstances and cycles that prompted the development of divine bodies. By concentrating on the geography, organization, and climates of these universes, researchers can sort out the multifaceted story of our planetary group's set of experiences.

Mars, frequently thought to be the most Earth-like planet, has been a point of convergence of investigation because of its true capacity as a host for past or present life. The quest for proof of microbial life on Mars resolves the subject of extraterrestrial life as well as gives experiences into the circumstances vital for life to emerge. Automated missions, for example, the Mars meanderers and orbiters sent off on strong rockets, have uncovered enticing signs about the planet's watery past and the chance of subsurface territories that might hold onto life.

Venus, Earth's singing twin, presents an unmistakable difference to the circumstances helpful for life. The investigation of Venus, worked with by mechanical missions and shuttle sent off into space on rockets, adds to how we might interpret planetary development and the likely outcomes of out of control nursery impacts. Notwithstanding its unfriendly nature, Venus holds significant illustrations about the fragile equilibrium of barometrical circumstances that support life on The planet.

Past our nearby neighbors, the gas goliaths Jupiter and Saturn, with their companies of moons and mind boggling ring frameworks, give a brief look into the variety of planetary frameworks. The investigation of these goliaths, directed through missions like Juno and Cassini sent off on rockets, reveals insight into the instruments driving planetary airs, attractive fields, and complex moon frameworks. The investigation of Jupiter's moon Europa, for example, makes the way for exploring subsurface seas and the potential for extraterrestrial life.

The meaning of interplanetary investigation stretches out to the external scopes of our planetary group, where frigid universes like Uranus and Neptune entice for study. These ice monsters, investigated by space apparatus sent off on board rockets, offer bits of knowledge into the elements of planetary airs and the secrets of planetary development in the external planetary group. The automated investigation of these far off planets adds layers to how we might interpret the tremendous and interconnected heavenly dance that administers the elements of the whole planetary group.

Interplanetary investigation isn't exclusively about uncovering the land and barometrical mysteries of far off universes; it likewise assumes an essential part in propelling comprehension we might interpret planetary frameworks past our own. The disclosure of exoplanets, planets circling stars past our planetary group, has arisen as a prospering field in astronomy. The investigation of these far off universes, worked with by space telescopes sent off on rockets, adds to how we might interpret planetary variety, tenability, and the circumstances important for life to exist somewhere else in the cosmic system.

The mechanical headways driven by interplanetary investigation have sweeping ramifications for space go and our capacity to wander past our planetary group. Rockets, the workhorses of room investigation, have developed from early fluid filled plans to modern and strong vehicles equipped for impelling space apparatus into the most distant compasses of the universe. The improvement of cutting edge drive frameworks, life support advancements, and independent route frameworks, all prodded by the difficulties of interplanetary missions, straightforwardly impacts our ability to investigate other star frameworks later on.

Additionally, the examples gained from interplanetary investigation have direct applications to challenges on The planet. The advancement of life emotionally supportive networks, shut circle ecological control, and asset use innovations for long-length space missions has suggestions for reasonable residing on our home planet. The inventive arrangements conceived for space head out can possibly resolve issues connected with ecological protection, asset shortage, and catastrophe reaction on The planet.

The quest for extraterrestrial knowledge (SETI) is one more feature of interplanetary investigation that reverberates with the human craving to grasp our spot in the universe. The organization of radio telescopes and the investigation of transmissions from far off stars, frequently sent off into space on strong rockets, address our journey to track down indications of astute life past Earth. While SETI has not yet identified definitive proof of extraterrestrial signals, the hunt proceeds, powered by the conviction that the revelation of extraterrestrial knowledge would upset how we might interpret the universe and our place in it.

Interplanetary investigation, with its innate difficulties and wins, moves a feeling of marvel and stunningness that rises above public lines and social limits. The famous pictures caught by space apparatus sent off on rockets, from the peaceful scenes of Mars to the great rings of

Saturn, have the ability to summon an aggregate appreciation for the magnificence and intricacy of the universe. These pictures serve as logical information as well as a wellspring of motivation that cultivates a common feeling of humankind's association with the more extensive universe.

The cooperative idea of interplanetary investigation mirrors the aggregate goals of humankind to arrive at past the bounds of our home planet. Worldwide associations, exemplified by missions like the Global Space Station (ISS) and cooperative endeavors between space organizations, exhibit the potential for participation chasing logical information. Rockets become the vessels that convey the yearnings of different countries into the enormous boondocks, advancing a feeling of solidarity in the investigation of the unexplored world.

As we ponder the meaning of interplanetary investigation, it becomes clear that this try is in excess of a logical venture — it is a demonstration of human interest, versatility, and the quest for information. The send off of rockets conveying automated messengers to far off planets encapsulates our inborn longing to investigate, to push the limits of what is known, and to look for replies to mature old inquiries concerning our starting points and the presence of life past Earth.

All in all, interplanetary investigation, made conceivable by the starting of space apparatus into space on strong rockets, addresses a groundbreaking excursion into the universe. It goes past the logical mission for information, diving into the domains of reasoning, motivation, and our common human story. The investigation of planets, moons, and divine bodies in our planetary group and past grows how we might interpret the universe as well as pushes us toward new outskirts of revelation and investigation. Rockets, as the empowering influences of this fabulous odyssey, convey with them the expectations and desires of humankind as we proceed with our grandiose journey into the unexplored world.

Investigation, both on The planet and then some, has been a principal part of human instinct since our earliest days. As we shift focus over to the universe, the demonstration of investigating the huge scopes of the universe takes on new aspects, turning into an excursion into the obscure that propels our logical comprehension as well as takes advantage of the center of our inborn interest and mission for information.

At the core of inestimable investigation lies the craving to appreciate the beginnings and nature of the universe. The universe, with its worlds, stars, planets, and infinite peculiarities, presents an embroidery of colossal intricacy and excellence. From the perspective of investigation, we try to unwind the secrets of the universe, looking into the profundities of existence to comprehend the essential powers that oversee the universe's development.

The investigation of space starts with our own heavenly area — the nearby planet group. From the internal rough planets to the external gas monsters, every individual from our sun based family offers special experiences into the cycles that formed our inestimable home.

Rockets, as the vehicles of our enormous yearnings, assume an essential part in sending off shuttle on missions to investigate these far off universes.

Mars, frequently alluded to as the "Red Planet," has enthralled human creative mind for a really long time. Mechanical missions, sent off on strong rockets, have dared to Mars to concentrate on its surface, environment, and potential for past or present life. The progress of missions like the Mars meanderers — Soul, Opportunity, Interest, and Determination — features the headways in rocket innovation that empower us to cross the Martian landscape, dissect its geography, and quest for indications of tenability.

Venus, Earth's nearest planetary neighbor, presents a differentiating universe of outrageous temperatures and a thick, destructive climate. Mechanical missions, including orbiters and landers sent off into space on rockets, have tested the secrets of Venus. The bits of knowledge acquired contribute not exclusively to figuring out the planet's advancement yet additionally offer similar information for Earth's own air and geographical cycles.

The external planets, Jupiter and Saturn, with their broad frameworks of moons and dazzling ring frameworks, coax for investigation. Space apparatus sent off on board rockets, like the Juno mission to Jupiter and the Cassini-Huygens mission to Saturn, have given phenomenal perspectives and information about these gas monsters. The investigation of their moons, including Europa and Titan, discloses different scenes and potential conditions that flash conversations about the chance of extraterrestrial life.

Interplanetary investigation stretches out past our planetary group to far off stars and exoplanets. Telescopes sent off into space on rockets, for example, the Hubble Space Telescope and the Kepler Space Telescope, have widened how we might interpret the universe by distinguishing exoplanets in the livable zones of other star frameworks. The quest for Earth-like planets and expected indications of something going on under the surface past our planetary group mirrors the extension of our infinite interest.

The mechanical wonders empowering these investigations are appeared in rockets — strong vehicles intended to defeat Earth's gravitational draw and transport payloads into space. The development of rocket innovation, from early fluid powered plans to current send off vehicles, marks achievements in human accomplishment. Rockets have turned into the passages to the universe, working with our excursion into space and opening the insider facts of the universe.

As we investigate the universe, the idea of room travel and colonization turns into a captivating wilderness. The Moon, Earth's normal satellite, has been the subject of both logical review and goals for human settlement. Rockets, conveying space apparatus and possible lunar environments, act as the channels for mankind to lay out a presence past Earth. The Artemis program, drove by NASA, epitomizes the cooperative work to return people to the Moon and lay the basis for future interplanetary investigation.

Mars, with its true capacity for human residence, addresses a drawn out objective for space offices and confidential endeavors. Rockets, as the method for transport, become the key part for ran missions to the Red Planet. The Space Send off Framework (SLS) and other send off vehicles are vital in understanding the fantasy about sending people to Mars, beating the difficulties of interplanetary travel and supporting life in the unforgiving Martian climate.

The meaning of vast investigation stretches out past logical undertakings — it rises above into the domains of motivation, creative mind, and the human soul. The pictures caught by space tests and telescopes sent off into space on rockets inspire a feeling of marvel and wonder. From the Mainstays of Creation in the Hawk Cloud to the peaceful magnificence of Saturn's rings, these visuals rouse thought about the endlessness and excellence of the universe.

Thoughtfully, vast investigation brings up significant issues about our spot in the universe. As rockets move shuttle into space, conveying our logical instruments and our expectations for interplanetary disclosure, they represent our aggregate process into the unexplored world. The investigation of the universe turns into a common undertaking, a demonstration of mankind's journey for information and grasping, rising above public limits and social contrasts.

Man-made reasoning (computer based intelligence) has turned into a necessary piece of inestimable investigation, improving the capacities of shuttle and telescopes sent off into space on rockets. Man-made intelligence calculations empower independent route, information investigation, and ongoing decision-production during space missions. AI models process immense datasets, permitting space apparatus to adjust to unanticipated difficulties and settle on choices in complex conditions. The cooperative energy among rockets and man-made intelligence addresses an innovative marriage that intensifies our capacity to investigate the universe.

The lawful and moral elements of infinite investigation are brought to the front as we adventure farther into space. The Space Deal, endorsed in 1967, structures the underpinning of global space regulation. It underscores the serene utilization of space, precludes the militarization of divine bodies, and restricts cases of public power. As privately owned businesses become progressively engaged with space exercises, inquiries of asset possession, ecological effect, and space administration come to the cutting edge. A cooperative and comprehensive way to deal with space regulation is critical to guarantee mindful and manageable investigation past Earth.

Public commitment and instructive effort assume critical parts in the outcome of enormous investigation. The amazing send-offs of rockets, the charming pictures from space missions, and the logical disclosures made in the universe catch the creative mind of individuals around the world. The capacity to share the marvels of room investigation encourages a worldwide appreciation for the significance of logical revelation and moves the up and coming age of researchers, designers, and pilgrims.

9

Chapter 8

Private Space Companies and the New Space Age

The beginning of the 21st century has seen a change in perspective in space investigation, set apart by the ascent of private space organizations that have impelled humankind into another space age. Customarily, space investigation was the elite area of government space organizations, with NASA in the US and Roscosmos in Russia driving the charge. Be that as it may, the scene has changed emphatically with the rise of private elements like SpaceX, Blue Beginning, and endless others. These organizations have infused another essentialness into space investigation, pushing limits, and rethinking what was once viewed as the last wilderness.

One of the pioneers in this new period of room investigation is SpaceX, established by business person Elon Musk in 2002. Musk's vision was aggressive yet clear: to decrease space transportation expenses and make space travel more open. SpaceX's lead project, the Bird of prey 1 rocket, accomplished circle in 2008, denoting whenever a secretly evolved fluid filled rocket first had arrived at Earth circle. This achievement was a harbinger of the groundbreaking effect privately owned businesses could have on space investigation.

The Bird of prey 1 achievement set up for SpaceX's resulting accomplishments, including the advancement of the Hawk 9 and Bird of prey Weighty rockets. Prominently, the Bird of prey 9 turned into the primary orbital-class rocket fit for reflight, exhibiting the plausibility of reusable rocket innovation. This cutting edge fundamentally brought down the expense of sending off payloads into space, cultivating a more practical and financially savvy way to deal with space investigation.

SpaceX's highest accomplishment came in 2020 with the effective maintained mission, Demo-2, which shipped NASA space explorers Douglas Hurley and Robert Behnken to the Global Space Station (ISS). This noticeable the initial time a business space apparatus, the Group Winged serpent, carried space explorers to the ISS. The mission highlighted the essential job privately owned businesses play in propelling human space investigation and catalyzed a recharged revenue in maintained spaceflight.

While SpaceX has been at the very front of the confidential space race, different organizations have likewise taken huge steps. Blue Beginning, established by Amazon's Jeff Bezos in 2000, has zeroed in on creating reusable rocket innovation and suborbital space the travel industry. Blue Beginning's New Shepard rocket, intended for suborbital flights, means to take paying clients on brief excursions to the edge of room. This try addresses a shift from the customary government-driven way to deal with space investigation towards an additional business and comprehensive model.

Moreover, a horde of more modest confidential space organizations has arisen, each adding to the enhancement and development of the space business. Rocket Lab, for example, has earned respect for its little satellite send off administrations, giving practical admittance to space for a developing number of clients. The expansion of private space organizations has democratized admittance to space, empowering colleges, research foundations, and more modest nations to take part in space investigation and satellite sending.

The ascent of private space organizations has not been restricted to Earth's circle. Aggressive designs to investigate and take advantage of assets from divine bodies have been gotten under way. Organizations like Planetary Assets and Profound Space Ventures mean to dig space rocks for important assets, like valuable metals and water. The expected overflow of assets in space could change businesses on The planet and act as an impetus for additional space investigation.

Past Earth's circle, the possibility of human settlement on different planets has turned into an unmistakable objective. SpaceX's Starship project, imagined as a completely reusable rocket equipped for conveying people to the Moon, Mars, and then some, addresses a striking step towards interplanetary travel. The fantasy about laying out a human presence on Mars, once consigned to the domain of sci-fi, is currently inside the domain of plausibility, because of the endeavors of private space organizations.

The new space age isn't exclusively characterized by the undertakings of American organizations. Worldwide, nations are perceiving the significance of cultivating a dynamic space industry, with a few countries effectively reassuring confidential area interest. India, through the Indian Space Exploration Association (ISRO), has embraced coordinated effort with privately owned businesses to support its space abilities. The outcome of ISRO's Polar Satellite Send off Vehicle (PSLV) in sending off little satellites has drawn in global business interest, further exhibiting the potential for coordinated effort between government organizations and confidential endeavors.

In Europe, the European Space Organization (ESA) has additionally embraced the changing scene by encouraging associations with private substances. The appearance of organizations like Arianespace has increased Europe's send off capacities, giving business send off administrations and adding to the worldwide space environment.

China, with its thriving space program, has likewise seen the rise of private space organizations. While China's space exercises are fundamentally supervised by government elements like the China Public

Space Organization (CNSA), privately owned businesses, for example, iSpace and OneWebChina have entered the scene, planning to cut out a specialty in the quickly growing space industry.

The union of government-upheld drives and confidential undertaking is molding another period of joint effort in space investigation. Organizations among public and confidential substances can possibly speed up progress, lessen costs, and widen the extent of room missions. The Worldwide Space Station fills in as a perfect representation of effective coordinated effort, with various nations contributing modules and assets to make a cooperative space natural surroundings.

As the confidential space area picks up speed, administrative systems should develop to oblige the changing elements of room investigation. Legislatures all over the planet are wrestling with the need to figure out some kind of harmony between cultivating development and guaranteeing security and natural obligation. The commercialization of room raises moral contemplations, including inquiries regarding asset abuse, ecological effect, and the expected militarization of room.

The administrative scene is additionally confounded by the worldwide idea of room exercises. The Space Settlement, endorsed by north of 100 nations, gives a basic structure to space investigation, underscoring the tranquil utilization of space and restricting the position of weapons of mass obliteration in circle. Nonetheless, the arrangement doesn't completely address the commercialization of room or the exercises of private elements.

Endeavors are in progress to foster global standards and rules for the dependable utilization of room. The Artemis Accords, drove by NASA and supported by a few nations, frame standards for lunar investigation, including the manageable and serene utilization of lunar assets. While these agreements address a positive step towards global collaboration, challenges stay in accomplishing a brought together way to deal with managing private space exercises.

The commercialization of room has changed the elements of room investigation as well as led to a thriving space the travel industry.

Organizations like Virgin Cosmic and Blue Beginning are competing to make space the travel industry a reality for private people. Suborbital flights, offering a couple of moments of weightlessness and stunning perspectives on Earth from the edge of room, are ready to turn into the following outskirts in extravagance travel.

The development of room the travel industry brings up issues about availability, wellbeing, and the ecological effect of business space travel. While space the travel industry stays a world class insight because of its significant expenses, progressing advancements in innovation and foundation might prepare for more extensive public cooperation later on. Adjusting the charm of room the travel industry with moral and ecological contemplations will be essential as this early industry advances.

The commercialization of room has additionally prodded advancement in satellite innovation, correspondence frameworks, and Earth perception capacities. Little satellites, frequently alluded to as Cube-Sats, are assuming a critical part in propelling space-based applications. These small satellites, with sizes going from a couple of centimeters to a couple of kilograms, are savvy and flexible, making them ideal for various missions, including logical exploration, Earth perception, and media communications.

Progressions in satellite innovation have prompted a multiplication of satellite heavenly bodies, with organizations like SpaceX's Starlink, OneWeb, and Amazon's Venture Kuiper wanting to send enormous organizations of satellites to give worldwide broadband web inclusion. While these drives hold the commitment of associating underserved districts and spanning the computerized partition, concerns have been raised about the potential for space garbage, impedance with galactic perceptions, and the militarization of room for correspondence purposes.

The rising clog in Earth's circle and the developing danger of room flotsam and jetsam have provoked calls for upgraded space traffic the executives and trash relief measures. As additional satellites are sent off into space, the gamble of impacts and the age of room flotsam and jetsam increments. Endeavors to foster global rules for capable space conduct and the execution of crash evasion frameworks are basic to guaranteeing the drawn out manageability of room exercises.

The approach of private space organizations has additionally powered headways in space investigation past our planetary group. Advancement drives, for example, the Leading edge Starshot project, mean to send little, lightweight space apparatus to our closest star framework, Alpha Centauri, for a portion of the speed of light. The interstellar investigation project imagines utilizing strong lasers to push nanocraft, outfitted with scaled down sensors, towards the far off star framework, possibly changing comprehension we might interpret the universe.

In the journey for interstellar investigation, confidential drives supplement customary government-drove endeavors, displaying the variety of approaches and the cooperative soul that characterizes the new space age. The quest for extraterrestrial life and the investigation of far off exoplanets have become key targets, driven by both logical interest and the potential for finding livable universes past our nearby planet group.

The advancement of private space organizations has not been without challenges. Specialized misfortunes, monetary limitations, and administrative obstacles have tried the strength of these undertakings. The inborn dangers related with space investigation, from send off disappointments to in-circle abnormalities, highlight the intricacy of wandering past Earth's environment. In spite of these difficulties, the assurance and creativity of privately owned businesses have energized a feeling of development and investigation that keeps on driving the business forward.

The new space age is portrayed by accomplishments in space investigation as well as by the developing crossing point of room and different businesses. The commercialization of room has set out open doors for joint effort with areas like broadcast communications, horticulture, money, and medical care. Satellite-based advances, including Earth perception and remote detecting, assume a significant part in tending to worldwide difficulties, for example, environmental change, debacle reaction, and asset the board.

The mix of room advancements into regular daily existence is exemplified by the expansion of satellite route frameworks, ordinarily utilized in route applications and area based administrations. The Worldwide Situating Framework (GPS), at first created for military purposes, has turned into a crucial device for regular citizen applications, going from transportation and operations to farming and outside diversion.

As confidential space organizations keep on pushing the limits of investigation, the job of room in molding the eventual fate of humankind couldn't possibly be more significant. The new space age addresses a combination of logical revelation, mechanical development, and business venture, with significant ramifications for how we might interpret the universe and our spot in the universe.

The democratization of room, worked with by privately owned businesses, has introduced a period where the universe is as of now not the selective space of legislatures however a wilderness open to investigation by different substances. The opportunities for logical revelation, asset usage, and human settlement in space are growing, offering an enticing vision of a future where mankind turns into a multi-planetary animal varieties.

Be that as it may, with these valuable open doors come moral, legitimate, and ecological obligations. The requirement for worldwide coordinated effort and the improvement of manageable practices in space exercises are basic to guarantee the drawn out suitability of our undertakings past Earth. As confidential space organizations keep on molding the direction of the new space age, the aggregate endeavors of states, industry partners, and the worldwide local area will assume an essential part in exploring the difficulties and valuable open doors that lie ahead.

8.1 Rise of private companies in the space industry

The 21st century has seen a seismic change in the scene of room investigation, set apart by the ascendance of privately owned businesses in the space business. Generally, space investigation had been the elite area of government space organizations like NASA, ESA, and Roscosmos. Be that as it may, the approach of privately owned businesses, drove by visionary business people and filled by mechanical advancement, has on a very basic level changed the elements of the space area.

Perhaps of the most conspicuous pioneer in this new time is SpaceX, established by business visionary Elon Musk in 2002. Musk's vision was to alter space travel by diminishing expenses and making it more open. The early long stretches of SpaceX were portrayed by aggressive objectives and critical difficulties. The organization's advancement came in 2008 when the Hawk 1 turned into the primary secretly evolved fluid filled rocket to arrive at Earth circle. This achievement denoted the start of another section in space investigation, one where privately owned businesses could contend on the worldwide stage.

Ensuing achievements hardened SpaceX's situation as a vital participant in the space business. The Bird of prey 9, a two-stage rocket intended for the solid and safe vehicle of satellites and the Mythical serpent space apparatus into space, turned into a workhorse for business and government dispatches. Be that as it may, the genuine huge advantage was the improvement of reusable rocket innovation. SpaceX showed the possibility of reusing rocket parts, fundamentally bringing down send off expenses and preparing for another period of financially savvy space travel.

In 2012, SpaceX left a mark on the world by turning into the main secretly financed organization to send a rocket, the Mythical serpent, to the Global Space Station (ISS). This obvious a defining moment, as it exhibited that privately owned businesses couldn't fabricate dependable rockets yet in addition add to the basic undertaking of resupplying and supporting the ISS.

The next years saw a progression of effective send-offs, including the presentation of the Hawk Weighty, the most remarkable functional rocket on the planet.

Notwithstanding, SpaceX's most huge accomplishment according to many was the Group Mythical serpent's effective manned mission, Demo-2, in 2020. This noticeable whenever a business rocket first moved space explorers to the ISS. It was a noteworthy second that highlighted the capacity of privately owned businesses to embrace maintained space missions, a space customarily held for government organizations. This accomplishment exhibited the specialized ability of SpaceX as well as shown the potential for business substances to assume a main part in human space investigation.

While SpaceX has been the flagbearer of the confidential space industry, different organizations have additionally made significant commitments. Blue Beginning, established by Amazon's Jeff Bezos in 2000, entered the scene with an emphasis on reusable rocket innovation and suborbital space the travel industry. Blue Beginning's New Shepard rocket, intended for suborbital flights, expects to offer paying clients a short excursion to the edge of room. This try addresses a shift from conventional government-driven ways to deal with a more business and comprehensive model of room investigation.

Notwithstanding the deeply grounded players, a large number of more modest confidential space organizations have arisen, each adding to the enhancement and development of the space business. Rocket Lab, for example, has earned respect for its little satellite send off administrations, giving financially savvy admittance to space for a developing number of clients. The democratization of room access has empowered colleges, research establishments, and more modest nations to take part effectively in space investigation and satellite sending.

The ascent of private space organizations isn't restricted to the US. Worldwide, nations are perceiving the capability of private area contribution in space exercises. In India, the Indian Space Exploration Association (ISRO) has embraced joint effort with privately owned businesses to reinforce its space capacities. The progress of ISRO's Polar Satellite Send off Vehicle (PSLV) in sending off little satellites has drawn in global business interest, displaying the potential for joint effort between government organizations and confidential ventures.

Likewise, in Europe, the European Space Office (ESA) has adjusted to the changing scene by cultivating associations with private elements. Arianespace, an European send off specialist co-op, plays had an essential impact in expanding Europe's send off capacities, offering business send off administrations and adding to the worldwide space biological system.

China, with its quickly propelling space program, has additionally seen the rise of private space organizations. While the essential space exercises in China are supervised by government substances like the China Public Space Organization (CNSA), privately owned businesses like iSpace and OneWebChina have entered the scene, planning to cut out a specialty in the quickly extending space industry.

The combination of government-supported drives and confidential undertaking is molding another time of cooperation in space investigation. Organizations among public and confidential elements can possibly speed up progress, diminish costs, and expand the extent of room missions. The Global Space Station fills in as a perfect representation of effective cooperation, with numerous nations contributing modules and assets to make a cooperative space environment.

As the confidential space area picks up speed, administrative structures should develop to oblige the changing elements of room investigation. Legislatures all over the planet are wrestling with the need to work out some kind of harmony between encouraging advancement and guaranteeing security and ecological obligation. The commercialization of room raises moral contemplations, including inquiries regarding asset double-dealing, natural effect, and the likely militarization of room.

The administrative scene is additionally convoluted by the worldwide idea of room exercises. The Space Settlement, endorsed by north of 100 nations, gives a primary system to space investigation, underlining the quiet utilization of space and disallowing the position of weapons of mass obliteration in circle. In any case, the settlement doesn't extensively address the commercialization of room or the exercises of private elements.

Endeavors are in progress to foster worldwide standards and rules for the dependable utilization of room. The Artemis Accords, drove by NASA and supported by a few nations, frame standards for lunar investigation, including the economical and tranquil utilization of lunar assets. While these agreements address a positive step towards global participation, challenges stay in accomplishing a brought together way to deal with managing private space exercises.

The commercialization of room has changed the elements of room investigation as well as led to a blossoming space the travel industry. Organizations like Virgin Cosmic and Blue Beginning are competing to make space the travel industry a reality for private people. Suborbital flights, offering a couple of moments of weightlessness and stunning perspectives on Earth from the edge of room, are ready to turn into the following outskirts in extravagance travel.

The development of room the travel industry brings up issues about availability, wellbeing, and the natural effect of business space travel. While space the travel industry stays a first class insight because of its significant expenses, progressing advancements in innovation and framework might prepare for more extensive public support from now on. Adjusting the appeal of room the travel industry with moral and natural contemplations will be urgent as this early industry develops.

The commercialization of room has likewise prodded development in satellite innovation, correspondence frameworks, and Earth perception capacities. Little satellites, frequently alluded to as CubeSats, are assuming an essential part in propelling space-based applications. These smaller than normal satellites, with sizes going from a couple of centimeters to a couple of kilograms, are savvy and flexible, making them ideal for different missions, including logical examination, Earth perception, and broadcast communications.

Progressions in satellite innovation have prompted a multiplication of satellite heavenly bodies, with organizations like SpaceX's Starlink,

OneWeb, and Amazon's Task Kuiper wanting to convey huge organizations of satellites to give worldwide broadband web inclusion. While these drives hold the commitment of associating underserved locales and connecting the computerized partition, concerns have been raised about the potential for space trash, obstruction with galactic perceptions, and the militarization of room for correspondence purposes.

The rising clog in Earth's circle and the developing danger of room garbage have provoked calls for improved space traffic the executives and flotsam and jetsam moderation measures. As additional satellites are sent off into space, the gamble of impacts and the age of room flotsam and jetsam increments. Endeavors to foster worldwide rules for capable space conduct and the execution of impact aversion frameworks are basic to guaranteeing the drawn out maintainability of room exercises.

The appearance of private space organizations has additionally filled progressions in space investigation past our nearby planet group. Advancement drives, for example, the Leading edge Starshot project, mean to send little, lightweight shuttle to our closest star framework, Alpha Centauri, for a portion of the speed of light. The interstellar investigation project imagines utilizing strong lasers to push nanocraft, outfitted with scaled down sensors, towards the far off star framework, possibly altering how we might interpret the universe.

In the journey for interstellar investigation, confidential drives supplement conventional government-drove endeavors, exhibiting the variety of approaches and the cooperative soul that characterizes the new space age. The quest for extraterrestrial life and the investigation of far off exoplanets have become key targets, driven by both logical interest and the potential for finding tenable universes past our planetary group.

The development of private space organizations has not been without challenges. Specialized difficulties, monetary imperatives, and administrative obstacles have tried the flexibility of these ventures. The inborn dangers related with space investigation, from send off disappointments to in-circle abnormalities, highlight the intricacy of wandering past Earth's climate. Notwithstanding these difficulties, the assurance and creativity of privately owned businesses have filled a feeling of development and investigation that keeps on driving the business forward.

The new space age is described by accomplishments in space investigation as well as by the developing crossing point of room and different enterprises. The commercialization of room has set out open doors for cooperation with areas like broadcast communications, agribusiness, money, and medical services. Satellite-based advancements, including Earth perception and remote detecting, assume a pivotal part in tending to worldwide difficulties, for example, environmental change, fiasco reaction, and asset the board.

The joining of room advancements into day to day existence is exemplified by the multiplication of satellite route frameworks, regularly utilized in route applications and area based administrations. The Worldwide Situating Framework (GPS), at first created for military purposes, has turned into an irreplaceable device for regular citizen applications, going from transportation and planned operations to horticulture and outside amusement.

As confidential space organizations keep on pushing the limits of investigation, the job of room in forming the fate of humankind couldn't possibly be more significant. The new space age addresses a combination of logical revelation, mechanical development, and business venture, with significant ramifications for how we might interpret the universe and our position in the universe.

The democratization of room, worked with by privately owned businesses, has introduced a period where the universe is as of now not the select space of states however an outskirts open to investigation by different substances. The opportunities for logical revelation, asset use, and human settlement in space are extending, offering a tempting vision of a future where mankind turns into a multi-planetary animal varieties.

Notwithstanding, with these potential open doors come moral, lawful, and ecological obligations. The requirement for worldwide cooperation and the advancement of feasible practices in space exercises are basic to guarantee the drawn out suitability of our undertakings past Earth. As confidential space organizations keep on forming the direction of the new space age, the aggregate endeavors of states, industry partners, and the worldwide local area will assume a critical part in exploring the difficulties and valuable open doors that lie ahead.

8.2 SpaceX, Blue Origin, and other key players

The new space age, set apart by the noticeable quality of privately owned businesses, is characterized by a small bunch of central participants that have reformed the scene of room investigation. Among these, SpaceX stands apart as a spearheading force, changing the business with its visionary objectives and notable accomplishments. Established in 2002 by business person Elon Musk, SpaceX, or Space Investigation Advances Corp., immediately became inseparable from development, cost-viability, and reusability in space transportation.

One of SpaceX's initial victories was the Hawk 1, which, in 2008, turned into the main secretly evolved fluid powered rocket to arrive at Earth circle. This achievement showed the way that privately owned businesses could contend in the domain of room investigation, generally overwhelmed by government organizations. Following this achievement, SpaceX fostered the Hawk 9, a flexible two-stage rocket intended for solid and safe vehicle of satellites and the Winged serpent space apparatus into space.

A characterizing element of SpaceX's methodology is the quest for reusable rocket innovation. The Hawk 9 turned into the main orbital-class rocket equipped for reflight, essentially bringing down send off costs by permitting the reuse of significant parts. This advancement has reshaped the financial matters of room travel, making it more practical and manageable. The progress of reusability arrived at its pinnacle with the Hawk Weighty, as of now the most remarkable functional rocket universally, equipped for lifting weighty payloads to circle.

Notwithstanding, SpaceX's highest accomplishment came in 2020 with the noteworthy Demo-2 mission. This mission denoted the main manned send off by a privately owned business, as NASA space explorers

Douglas Hurley and Robert Behnken were moved to the Global Space Station (ISS) on board the Team Winged serpent space apparatus. The outcome of Demo-2 approved SpaceX's capacities in ran spaceflight as well as flagged another time where business elements assumed a focal part in human space investigation.

Past Earth's circle, SpaceX has focused on aggressive interplanetary missions. The Starship project, as of now being developed, intends to be a completely reusable shuttle fit for conveying people to the Moon, Mars, and then some. The nervy objective of laying out a human presence on Mars addresses a change in perspective in our way to deal with space investigation, imagining a future where mankind turns into a multi-planetary animal types.

While SpaceX has been at the very front of the confidential space race, Blue Beginning, established by Amazon's Jeff Bezos in 2000, has likewise arisen as a critical player. Blue Beginning spotlights on creating reusable rocket innovation and suborbital space the travel industry. The New Shepard rocket, named after Alan Shepard, the main American in space, is intended for suborbital flights, fully intent on taking paying clients on brief excursions to the edge of room.

Blue Beginning's accentuation on reusability lines up with its drawn out vision of empowering a large number of individuals to live and work in space. The organization imagines a future where space travel isn't simply open to space travelers yet additionally to private people looking for the experience of weightlessness and all encompassing perspectives from the edge of Earth's environment. While Blue Beginning still can't seem to lead manned spaceflights for paying clients, its desires in space the travel industry address a clever way to deal with commercializing space exercises.

Notwithstanding SpaceX and Blue Beginning, a heap of more modest confidential space organizations has entered the field, each adding to the enhancement and extension of the space business. Rocket Lab, an organization established in 2006 by New Zealand business person Peter Beck, has earned respect for its emphasis on giving little satellite send off administrations. The Electron rocket, intended for conveying little payloads into space, has turned into a staple for savvy admittance to circle.

Rocket Lab's Electron rocket is outstanding for its committed spotlight on the little satellite market, offering regular and custom-made send off administrations for a developing number of clients. The scaling down of satellites, known as CubeSats, has flooded in fame, and Rocket Lab's capacity to take special care of this request has situated the organization as a central participant in the quickly developing space industry.

The ascent of private space organizations isn't select to the US. Globally, nations are perceiving the capability of private area association in space exercises. In India, the Indian Space Exploration Association (ISRO) has embraced coordinated effort with privately owned businesses to upgrade its space capacities. The outcome of ISRO's Polar Satellite Send off Vehicle (PSLV) in sending off little satellites has collected global business interest, exhibiting the potential for associations between government organizations and confidential undertakings.

In Europe, the European Space Organization (ESA) has adjusted to the changing scene by cultivating coordinated efforts with private substances. Arianespace, an European send off specialist organization, plays had a critical impact in expanding Europe's send off capacities, offering business send off administrations and adding to the worldwide space environment. The development of private players in Europe supplements conventional government-drove endeavors and adds to the general imperativeness of the European space area.

China, with its thriving space program, has likewise seen the development of private space organizations. While the essential space exercises in China are regulated by government substances like the China Public Space Organization (CNSA), privately owned businesses like iSpace and OneWebChina have entered the scene. These organizations mean to cut out a specialty in the quickly growing space industry, adding to the expansion of China's space capacities.

The coordinated effort between government-supported drives and confidential undertaking is molding another period of room investigation. Organizations among public and confidential elements can possibly speed up progress, diminish costs, and widen the extent of room missions. The Global Space Station (ISS) fills in as a perfect representation of effective joint effort, with various nations contributing modules and assets to make a cooperative space natural surroundings.

As the confidential space area picks up speed, administrative systems should develop to oblige the changing elements of room investigation. States overall are wrestling with the need to find some kind of harmony between encouraging development and guaranteeing wellbeing and natural obligation. The commercialization of room raises moral contemplations, including inquiries regarding asset double-dealing, ecological effect, and the likely militarization of room.

The administrative scene is additionally muddled by the worldwide idea of room exercises. The Space Settlement, endorsed by more than 100 nations, gives a fundamental structure to space investigation, underlining the tranquil utilization of space and disallowing the position of weapons of mass obliteration in circle. Nonetheless, the settlement doesn't extensively address the commercialization of room or the exercises of private substances.

Endeavors are in progress to foster worldwide standards and rules for the dependable utilization of room. The Artemis Accords, drove by NASA and supported by a few nations, frame standards for lunar investigation, including the economical and tranquil utilization of lunar assets. While these agreements address a positive step towards worldwide collaboration, challenges stay in accomplishing a bound together way to deal with controlling confidential space exercises.

The commercialization of room has changed the elements of room investigation as well as led to a prospering space the travel industry. Organizations like Virgin Cosmic, established by Sir Richard Branson, and Blue Beginning are competing to make space the travel industry a reality for private people. Suborbital flights, offering a couple of moments of weightlessness and stunning perspectives on Earth from the edge of room, are ready to turn into the following wilderness in extravagance travel.

The rise of room the travel industry brings up issues about availability, security, and the ecological effect of business space travel. While space the travel industry stays a tip top encounter because of its significant expenses, progressing improvements in innovation and foundation might prepare for more extensive public support from here on out. Adjusting the appeal of room the travel industry with moral and ecological contemplations will be critical as this early industry develops.

The commercialization of room has additionally prodded development in satellite innovation, correspondence frameworks, and Earth perception abilities. Little satellites, frequently alluded to as CubeSats, are assuming a significant part in propelling space-based applications. These small scale satellites, with sizes going from a couple of centimeters to a couple of kilograms, are practical and flexible, making them ideal for various missions, including logical exploration, Earth perception, and broadcast communications.

Progressions in satellite innovation have prompted an expansion of satellite heavenly bodies, with organizations like SpaceX's Starlink, OneWeb, and Amazon's Task Kuiper wanting to convey huge organizations of satellites to give worldwide broadband web inclusion. While these drives hold the commitment of associating underserved districts and spanning the advanced gap, concerns have been raised about the potential for space garbage, obstruction with galactic perceptions, and the militarization of room for correspondence purposes.

The rising clog in Earth's circle and the developing danger of room flotsam and jetsam have provoked calls for upgraded space traffic the board and trash alleviation measures. As additional satellites are sent off into space, the gamble of crashes and the age of room trash increments. Endeavors to foster global rules for dependable space conduct and the execution of impact aversion frameworks are basic to guaranteeing the drawn out maintainability of room exercises.

The approach of private space organizations has likewise powered headways in space investigation past our nearby planet group. Advancement drives, for example, the Cutting edge Starshot project, expect to send little, lightweight shuttle to our closest star framework, Alpha Centauri, for a portion of the speed of light. The interstellar investigation project imagines utilizing strong lasers to drive nanocraft, furnished with scaled down sensors, towards the far off star framework, possibly upsetting comprehension we might interpret the universe.

In the journey for interstellar investigation, confidential drives supplement customary government-drove endeavors, displaying the variety of approaches and the cooperative soul that characterizes the new space age. The quest for extraterrestrial life and the investigation of far off exoplanets have become key goals, driven by both logical interest and the potential for finding tenable universes past our planetary group.

The development of private space organizations has not been without challenges. Specialized misfortunes, monetary requirements, and administrative obstacles have tried the flexibility of these endeavors. The innate dangers related with space investigation, from send off disappointments to in-circle oddities, highlight the intricacy of wandering past Earth's environment. In spite of these difficulties, the assurance and resourcefulness of privately owned businesses have filled a feeling of development and investigation that keeps on driving the business forward.

The new space age is portrayed by accomplishments in space investigation as well as by the developing convergence of room and different enterprises. The commercialization of room has set out open doors for joint effort with areas like media communications, farming, money, and medical care. Satellite-based advances, including Earth perception and remote detecting, assume a pivotal part in tending to worldwide difficulties, for example, environmental change, calamity reaction, and asset the board.

The combination of room advances into regular daily existence is exemplified by the expansion of satellite route frameworks, ordinarily utilized in route applications and area based administrations. The Worldwide Situating Framework (GPS), at first produced for military purposes, has turned into an irreplaceable device for regular citizen applications, going from transportation and coordinated operations to horticulture and outside entertainment.

As confidential space organizations keep on pushing the limits of investigation, the job of room in molding the fate of mankind couldn't possibly be more significant. The new space age addresses a combination of logical revelation, mechanical development, and business undertaking, with significant ramifications for how we might interpret the universe and our spot in the universe.

The democratization of room, worked with by privately owned businesses, has introduced a period where the universe is as of now not the selective space of states however an outskirts open to investigation by different substances. The opportunities for logical disclosure, asset use, and human settlement in space are extending, offering an enticing vision of a future where mankind turns into a multi-planetary animal categories.

Nonetheless, with these open doors come moral, legitimate, and ecological obligations. The requirement for global coordinated effort and the advancement of feasible practices in space exercises are basic to guarantee the drawn out reasonability of our undertakings past Earth. As confidential space organizations keep on forming the direction of the new space age, the aggregate endeavors of legislatures, industry partners, and the global local area will assume an essential part in exploring the difficulties and open doors that lie ahead.

8.3 The impact of commercial space ventures on the future of rocketry

The scene of rocketry is going through a groundbreaking movement, driven by the rising impact of business space adventures. As the space business advances, pushed by the development and innovative soul of privately owned businesses, the effect on rocketry is significant, affecting innovation, availability, and the actual idea of room investigation.

One of the essential drivers of progress is SpaceX, established by Elon Musk in 2002. SpaceX's aggressive vision focuses on lessening the expense of room access and empowering human colonization of Mars. The organization's progressive methodology includes the improvement of reusable rocket innovation, on a very basic level changing the financial matters of room travel.

The Bird of prey 9, a workhorse in SpaceX's armada, is a two-stage rocket intended for dependability and cost-viability. The advancement lies in its capacity to land and be reused for resulting missions. The fruitful arrivals of Bird of prey 9 sponsors, both ashore and adrift, have shown the practicality of reusability, an idea recently thought to be unfeasible in the aeronautic trade. This advancement has sweeping ramifications, altogether diminishing send off expenses and making space investigation all the more monetarily practical.

SpaceX's quest for reusability arrived at new levels with the Bird of prey Weighty, the most remarkable functional rocket internationally. Including three Bird of prey 9 centers, the Hawk Weighty flaunts the capacity to lift weighty payloads into space. Its fruitful lady trip in 2018 denoted an achievement in the mission for strong yet savvy send off vehicles.

The effect of SpaceX's developments reaches out past expense decrease. The organization's Team Winged serpent shuttle, created to move space travelers to the Worldwide Space Station (ISS), represents the coordination of business adventures in manned space missions. The progress of the Demo-2 mission in 2020 denoted the main manned orbital send off by a privately owned business, highlighting the change in perspective in human spaceflight. Business substances are presently significant supporters of manned missions, a space customarily overwhelmed by government organizations.

SpaceX's aggressive Starship project further embodies the groundbreaking idea of business space adventures. Imagined as a completely reusable rocket, Starship expects to work with maintained missions to the Moon, Mars, and then some. In the event that effective, Starship could rethink the potential outcomes of profound space investigation, empowering mankind to turn into a multi-planetary animal varieties.

Past SpaceX, Blue Beginning, established by Jeff Bezos, has likewise assumed a huge part in molding the fate of rocketry. Blue Beginning's New Shepard rocket is intended for suborbital space the travel industry, determined to offer confidential people a concise excursion to the edge of room. While New Shepard right now centers around the travel industry, the reusable rocket innovation it utilizes adds to the more extensive pattern of making space travel more feasible and practical.

The effect of business space adventures on rocketry isn't restricted to the US. All around the world, nations are perceiving the capability of private area association in space exercises. In India, the Indian Space Exploration Association (ISRO) has cultivated coordinated efforts with privately owned businesses to upgrade its space abilities. The outcome of ISRO's Polar Satellite Send off Vehicle (PSLV) in sending off little satellites has drawn in worldwide business interest, exhibiting the potential for organizations between government offices and confidential ventures.

In Europe, the European Space Office (ESA) has adjusted to the changing scene by teaming up with private substances. Arianespace, an European send off specialist organization, plays had a urgent impact in expanding Europe's send off capacities, offering business send off administrations and adding to the worldwide space environment. The presence of private players in Europe supplements conventional government-drove endeavors, encouraging a different and cutthroat space industry.

China, with its quickly propelling space program, has additionally seen the rise of private space organizations. While government elements like the China Public Space Organization (CNSA) regulate essential space exercises, privately owned businesses, for example, iSpace and OneWebChina have entered the scene. These organizations expect to cut out a specialty in the quickly growing space industry, adding to the expansion of China's space capacities.

The joint effort between government-upheld drives and confidential venture is characterizing another period of room investigation. Organizations among public and confidential elements can possibly speed up progress, decrease costs, and expand the extent of room missions. The Global Space Station (ISS) fills in as a great representation of fruitful cooperation, with different nations contributing modules and assets to make a cooperative space environment.

As confidential space adventures pick up speed, administrative systems should develop to oblige the changing elements of room investigation. States overall are wrestling with the need to find some kind of harmony between cultivating development and guaranteeing security and natural obligation. The commercialization of room raises moral contemplations, including inquiries regarding asset double-dealing, ecological effect, and the likely militarization of room.

The administrative scene is additionally convoluted by the worldwide idea of room exercises. The Space Settlement, endorsed by more than 100 nations, gives a fundamental system to space investigation, stressing the tranquil utilization of space and forbidding the situation of weapons of mass obliteration in circle. In any case, the settlement doesn't completely address the commercialization of room or the exercises of private substances.

Endeavors are in progress to foster worldwide standards and rules for the mindful utilization of room. The Artemis Accords, drove by NASA and embraced by a few nations, frame standards for lunar investigation, including the practical and quiet utilization of lunar assets. While these agreements address a positive step towards global participation, challenges stay in accomplishing a bound together way to deal with controlling confidential space exercises.

The commercialization of room has changed the elements of room investigation as well as prodded advancement in satellite innovation, correspondence frameworks, and Earth perception capacities. Little satellites, frequently alluded to as CubeSats, are assuming a urgent part in propelling space-based applications. These smaller than normal satellites, with sizes going from a couple of centimeters to a couple of kilograms, are practical and flexible, making them ideal for various missions, including logical examination, Earth perception, and broadcast communications.

Progressions in satellite innovation have prompted an expansion of satellite heavenly bodies, with organizations like SpaceX's Starlink, OneWeb, and Amazon's Undertaking Kuiper wanting to send huge organizations of satellites to give worldwide broadband web inclusion.

While these drives hold the commitment of associating underserved locales and crossing over the computerized partition, concerns have been raised about the potential for space trash, impedance with galactic perceptions, and the militarization of room for correspondence purposes.

The rising clog in Earth's circle and the developing danger of room trash have provoked calls for upgraded space traffic the executives and flotsam and jetsam relief measures. As additional satellites are sent off into space, the gamble of crashes and the age of room garbage increments. Endeavors to foster global rules for mindful space conduct and the execution of impact aversion frameworks are basic to guaranteeing the drawn out maintainability of room exercises.

The approach of private space organizations has additionally filled progressions in space investigation past our nearby planet group. Advancement drives, for example, the Cutting edge Starshot project, plan to send little, lightweight space apparatus to our closest star framework, Alpha Centauri, for a portion of the speed of light. The interstellar investigation project imagines utilizing strong lasers to drive nanocraft, outfitted with scaled down sensors, towards the far off star framework, possibly altering how we might interpret the universe.

In the journey for interstellar investigation, confidential drives supplement conventional government-drove endeavors, displaying the variety of approaches and the cooperative soul that characterizes the new space age. The quest for extraterrestrial life and the investigation of far off exoplanets have become key targets, driven by both logical interest and the potential for finding tenable universes past our planetary group.

The advancement of private space organizations has not been without challenges. Specialized misfortunes, monetary requirements, and administrative obstacles have tried the versatility of these ventures. The intrinsic dangers related with space investigation, from send off disappointments to in-circle abnormalities, highlight the intricacy of wandering past Earth's environment. Regardless of these difficulties, the assurance and inventiveness of privately owned businesses have filled a feeling of development and investigation that keeps on driving the business forward.

The effect of business space adventures on the eventual fate of rocketry stretches out past the bounds of our planet. The prospering field of room the travel industry, pushed by organizations like Virgin Cosmic and Blue Beginning, looks to make space travel a reality for private people. Suborbital flights, offering a couple of moments of weightlessness and stunning perspectives on Earth from the edge of room, address the following outskirts in extravagance travel.

The rise of room the travel industry brings up issues about openness, wellbeing, and the natural effect of business space travel. While space the travel industry stays a world class insight because of its significant expenses, continuous advancements in innovation and framework might make ready for more extensive public cooperation later on. Adjusting the charm of room the travel industry with moral and ecological contemplations will be pivotal as this incipient industry develops.

Taking everything into account, the effect of business space adventures on the eventual fate of rocketry is both significant and complex. From the reusability insurgency spearheaded by SpaceX to the broadening of send off administrations by organizations like Rocket Lab, the scene of rocketry is advancing at a phenomenal speed. The coordinated effort among public and confidential elements, the quest for interplanetary investigation, and the democratization of room access all add to a future where space isn't simply the domain of states however an outskirts open to investigation by different substances.

As the new space age unfurls, the aggregate endeavors of states, industry partners, and the worldwide local area will be fundamental in exploring the difficulties and valuable open doors that lie ahead. The effect of business space adventures on the eventual fate of rocketry isn't simply a mechanical shift; it addresses a worldview change in our way to deal with space investigation. The combination of development, co-operation, and business undertaking is forming a future where human-kind's presence in space isn't just supported yet extended, proclaiming a strong and unfathomable period of investigation past Earth.

10

Chapter 9

Igniting the Future

In a world wavering near the very edge of exceptional mechanical progression, the blazes of development consume more splendid than any time in recent memory. Lighting what's in store isn't only an expression; it exemplifies the aggregate human undertaking to push the limits of what is conceivable. The persevering quest for information and progress impels social orders forward, molding the fate of ages yet unborn.

At the core of this journey for a more splendid tomorrow lies the steadily developing scene of innovation. The 21st century saw a dramatic flood in forward leaps, from man-made brainpower to biotechnology, every revelation filling in as a structure block for what's to come. The combination of disciplines and the intermingling of thoughts fuel a collaboration that rises above the limits of individual fields. It is inside this interdisciplinary pot that the sparkles of development track down ripe ground.

Man-made reasoning, once restricted to the domain of sci-fi, has turned into an essential piece of our regular routines. AI calculations examine immense datasets, uncovering examples and bits of knowledge that escape the human brain. Independent vehicles explore our roads, directed by calculations that advance as time passes. The cooperative connection among man and machine is as of now not a far off vision; it is our ongoing reality, a harbinger representing things to come.

Biotechnology, with its commitment of controlling the structure blocks of life, remains as a demonstration of human inventiveness. CRISPR-Cas9 innovation permits researchers to alter qualities with remarkable accuracy, making the way for another period of clinical conceivable outcomes. Hereditary treatments offer desire to those wrestling with hopeless sicknesses, introducing an age where customized medication turns into the standard instead of the exemption.

As the ringlets of innovation weave a mind boggling embroidery, the energy scene goes through its very own upheaval. Environmentally friendly power sources, once consigned to the edges of achievability, presently power countries and enlighten urban areas. Sun powered chargers tackle the unlimited energy of the sun, while wind turbines dance to the cadence of the always present breeze. The progress from petroleum derivatives to practical options denotes a urgent second in the continuous battle against environmental change, a fight battled on the bleeding edges of progress.

The computerized domain, an elusive space that rises above actual limits, is both a reflection and a harbinger of cultural change. The web, when a curiosity, has developed into a worldwide brain network interfacing minds across landmasses. Virtual entertainment stages act as computerized public square, where thoughts are traded, unrests are started, and the beat of shared awareness resonates. The democratization of data has enabled people, leading to another period of resident news-casting and activism.

However, with progress comes liability. The quick speed of innovative headway has exceeded the moral structures that ought to direct its direction. Inquiries of protection, security, and the possible abuse of strong advancements pose a potential threat not too far off. As we light the future, we should likewise fuel the flares of moral mindfulness, guaranteeing that the advantages of progress are fairly dispersed and that the potential damages are alleviated.

In the pot of advancement, the job of schooling couldn't possibly be more significant. What's in store has a place with those equipped with information, imagination, and flexibility. School systems should develop to encourage decisive reasoning, critical thinking abilities, and a profound comprehension of the moral ramifications of innovative progressions. Deep rooted learning becomes a mantra as well as a need in reality as we know it where the main steady is change.

The worldwide scene isn't invulnerable to the difficulties that go with progress. Disparities endure, both inside and among countries. The advantages of mechanical progression frequently accumulate lop-sidedly, augmenting the hole between the wealthy and the less wealthy. As we light the future, a pledge to inclusivity and civil rights should be woven into the texture of progress. The products of development ought not be the select space of a special minority however ought to elevate whole networks and countries.

In the domain of medical care, the combination of innovation and science proclaims another time of conceivable outcomes. Telemedicine carries medical services to the remotest corners of the globe, associating patients with specialists through the computerized ether. Wearable gadgets screen imperative signs progressively, giving a nonstop stream of information that enables people to assume responsibility for their wellbeing. The marriage of innovation and medical care isn't just about treating sicknesses however about advancing wellbeing and forestalling diseases before they flourish.

Space, the last wilderness, allures with the commitment of untold disclosures. The investigation of Mars, once restricted to the domain of sci-fi, is presently inside the grip of human accomplishment. Confidential endeavors, powered by the fantasies of visionaries, contend to cut their imprint on the universe. The start of rocket motors sends space apparatus rushing through the huge scope, opening new parts in the human odyssey.

As we put our focus on the stars, contemplating the ramifications of our excursion into the cosmos is fundamental. The illustrations gained from Earth — about ecological stewardship, participation, and the worth of life — should go with us on our interstellar journey. The start of rocket motors ought not be a takeoff from our obligations yet a continuation of our obligation to save and safeguard the delicate blue circle we call home.

Amidst this tireless walk forward, artistic expression and humanities stand as guides of contemplation and reflection. While science and innovation push the limits of what is conceivable, artistic expressions dive into the domains of creative mind and feeling. They help us to remember our common mankind, provoking us to consider what we can do as well as what we ought to do. The combination of the logical and the imaginative isn't a crash yet a dance, an amicable interchange that enhances the human experience.

What's in store is certainly not a far off objective; it is an excursion unfurling right now. Every choice we make, every revelation we celebrate, and each challenge we stand up to shapes the forms of tomorrow. The start representing things to come is certainly not a singular demonstration however an aggregate undertaking that rises above boundaries and ranges ages. In the pot of progress, we are not simple onlookers but rather dynamic members, employing the blazes of development with a significant consciousness of their power.

The difficulties that lie ahead are considerable, from environmental change and asset exhaustion to moral predicaments and international pressures. The start representing things to come requests a worldwide viewpoint, an acknowledgment that our interconnected world requires cooperative arrangements. The blazes of advancement should not be restricted to individual countries or partnerships but rather should be bridled to benefit mankind.

Chasing after progress, it is fundamental to consider the effect on the normal world that supports us. The start representing things to come ought not be a fire that consumes the actual underpinning of our reality. Economical practices, protection endeavors, and a profound regard for biodiversity should be basic to our methodology. The fire of progress ought to enlighten a way that prompts concordance with the climate, not conflict.

Moral contemplations stretch out past the bounds of innovation to envelop the actual texture of human cooperation. In our current reality where data streams unendingly, the obligation to disperse exact and unprejudiced information becomes principal. The start representing things to come is, in numerous ways, a fight for truth, a guarantee to maintaining the honesty of data notwithstanding deception and control.

The idea of administration goes through a change following mechanical progressions. As man-made reasoning additions conspicuousness, inquiries of responsibility, straightforwardness, and dynamic power become squeezing. The start representing things to come requests a reexamination of administration structures, guaranteeing that they develop to address the difficulties presented by a quickly changing innovative scene.

In the financial domain, the flares of progress should not stir up imbalance. The advantages of development ought to inspire whole social orders, giving open doors to all instead of a limited handful. The start representing things to come requires a monetary worldview that is comprehensive, maintainable, and aware of the social effect of innovative progressions.

As we explore the strange waters representing things to come, developing an outlook of flexibility and adaptability is urgent. The start of progress is frequently joined by choppiness and vulnerability. The capacity to weather conditions storms, gain from disappointments, and turn despite difficulties is a sign of social orders that flourish in the steadily changing scene representing things to come.

9.1 Current and upcoming developments in rocket technology

Rocket innovation, when the space of country states participated in international space races, has gone through an extraordinary development. In the ongoing time, a dynamic scene of public and confidential substances is effectively participated in propelling rocketry, proclaiming another period of room investigation and commercialization. As we dive into the current and forthcoming advancements in rocket innovation, it becomes clear that we stand at the cusp of phenomenal accomplishments that will shape the direction of human spacefaring tries.

One of the characterizing elements of contemporary rocket innovation is the rising pretended by confidential undertakings. Organizations like SpaceX, Blue Beginning, and Rocket Lab have arisen as vital participants in the space business, testing conventional models overwhelmed by government offices. SpaceX, established by business person Elon Musk, has gathered far and wide consideration for its aggressive objectives, including the advancement of reusable rocket parts and the colonization of Mars. The Hawk 9 rocket, with its reusable first stage, represents a change in perspective in cost-viability and manageability.

Reusability remains as a crucial development in rocket innovation, essentially decreasing the monetary boundaries related with space investigation. Customarily, rockets were single-use vehicles, with their parts disposed of after a solitary send off. The coming of reusable rocket stages, exemplified by SpaceX's Bird of prey 9 and Hawk Weighty, marks a takeoff from this show. The capacity to recuperate and revamp rocket parts brings down send off costs as well as lines up with more extensive maintainability objectives, diminishing the ecological effect of room exercises.

Past reusability, headways in drive frameworks assume a vital part in forming the scene of rocket innovation. Particle and electric impetus, which use electric fields to speed up particles and create push, are acquiring conspicuousness for specific mission profiles. While conventional compound rockets stay the workhorses for sending off payloads into space, electric impetus frameworks offer the upsides of eco-friendliness and delayed mission terms, making them ideal for profound space investigation.

In the journey for more noteworthy proficiency and capacity, new impetus advancements are under dynamic turn of events. Atomic warm drive, bridling the energy delivered by atomic responses to warm a fuel and produce push, holds the commitment of fundamentally diminishing travel times for maintained missions to far off objections. The utilization of atomic drive frameworks could upset human investigation of the Moon, Mars, and then some, opening the potential for supported presence in the external ranges of our planetary group.

In the domain of rocket plan, streamlined features and materials science assume critical parts in improving execution and versatility. Advancements in lightweight materials, like high level composites and compounds, add to the improvement of rockets with further developed payload limits and underlying honesty. Besides, progressions in computational liquid elements empower more precise recreations of a rocket's streamlined way of behaving, prompting upgraded plans that limit drag and improve generally execution.

Scaling down and modularization are patterns that have pervaded different businesses, and rocket innovation is no exemption. Little satellite send-offs, worked with by scaled down send off vehicles, offer savvy answers for conveying payloads into space. Organizations like Rocket Lab spend significant time in little satellite send-offs, offering committed types of assistance for a developing business sector of nanosatellites and microsatellites.

This pattern democratizes admittance to space, empowering scholastic establishments, new companies, and non-industrial countries to take part in space-based exercises.

As the limits of rocket innovation grow, so too do the desires for human investigation past Earth. The Moon, when the stage for verifiable Apollo missions, has recovered unmistakable quality as an objective for investigation and possible home. Government offices and privately owned businesses the same are creating plans for lunar missions, with an accentuation on laying out maintainable lunar bases. The Moon serves not just as a venturing stone for additional investigation yet in addition as a proving ground for advancements fundamental for future manned missions to Mars.

Mars, with its persona and potential for holding onto indications of past or present life, catches the aggregate creative mind of researchers, architects, and space aficionados. Plans for ran missions to Mars include unpredictable calculated difficulties, from life emotionally supportive networks to radiation insurance. The improvement of interplanetary transportation frameworks, like SpaceX's Starship, addresses a huge step towards making the fantasy of human colonization of Mars a substantial reality.

Notwithstanding attempts inside our planetary group, the quest for exoplanets and the potential for tenable universes past our heavenly area impel the mission for cutting edge impetus innovations. Ideas like laser drive and sun powered sails, which outfit outside energy hotspots for impetus, are under investigation for interstellar missions. While the difficulties of arriving at other star frameworks stay imposing, the quest for interstellar travel addresses a demonstration of human interest and the craving to investigate the enormous embroidery past recognizable skylines.

Worldwide joint effort is a basic part of the contemporary space age. The Worldwide Space Station (ISS), a cooperative exertion including space organizations from the US, Russia, Europe, Japan, and Canada, remains as an image of participation in space investigation. As the ISS proceeds with its central goal of logical examination and global coordinated effort, conversations about the foundation of lunar space stations and cooperative endeavors for Mars investigation gain unmistakable quality.

Progressions in rocket innovation are indistinguishable from the advancement of state of the art payload advancements. Earth perception satellites furnished with high-goal imaging capacities add to weather conditions checking, fiasco reaction, and natural administration. Correspondences satellites structure the foundation of worldwide network, working with consistent correspondence across mainlands. Logical payloads, going from space telescopes to planetary wanderers, extend how we might interpret the universe and our place inside it.

The intermingling of computerized reasoning (simulated intelligence) and rocket innovation addresses a boondocks that holds enormous potential. Computer based intelligence calculations streamline send off directions, foresee expected issues, and upgrade the independence of rocket. AI applications empower continuous information investigation and decision-production during missions, adding to the productivity and progress of room investigation attempts. The combination of computer based intelligence and rocket innovation epitomizes the collaboration between two groundbreaking fields.

Space trash, comprising of outdated satellites, spent rocket stages, and different parts, represents a developing test in Earth's circle. As the quantity of items in space builds, the gamble of crashes and the age of extra trash rise. Relieving space garbage requires worldwide participation and the improvement of systems for flotsam and jetsam expulsion and dependable space rehearses. Besides, the plan of future satellites and send off vehicles integrates contemplations for end-of-life removal, adding to the drawn out maintainability of room exercises.

In the space of rocket drive, green and practical choices are building up some decent momentum. Conventional rocket charges, depending on compound responses that discharge huge amounts of energy, raise ecological worries. Green fuels, using not so much harmful but rather more harmless to the ecosystem substances, offer a feasible other option. The quest for reasonable drive lines up with more extensive endeavors to diminish the natural effect of room exercises and advance capable space investigation.

The beginning field of room the travel industry addresses a change in perspective in space openness. Organizations like Blue Beginning and Virgin Cosmic plan to make suborbital spaceflight encounters accessible to private people. While space the travel industry is still in its beginning phases, the possibility of regular people wandering past Earth's environment mirrors a democratization of room access. The commercialization of room holds the possibility to finance further progressions in rocket innovation and space investigation.

The powerful scene of rocket innovation isn't without any trace of difficulties. Political, monetary, and administrative contemplations shape the direction of room projects and industry drives. Global coordinated efforts face international intricacies, and the distribution of assets for space investigation contends with other cultural needs. Finding some kind of harmony between public interests, confidential venture, and worldwide collaboration is a sensitive undertaking that requires keen discretionary and strategy choices.

9.2 The potential for space tourism and colonization

The possibility of room the travel industry and colonization remains as perhaps of the most dazzling and groundbreaking account in the continuous story of human investigation. While space investigation has generally been the space of legislatures and space organizations, the 21st century observes the beginning of another period where confidential undertakings and business adventures effectively partake in opening the space outskirts to regular folks.

This change in elements carries with it the commitment of encountering the miracles of room as well as laying out a human presence past Earth. As we investigate the potential for space the travel industry and colonization, we dig into the logical, mechanical, and moral aspects that shape this unfurling story.

Space the travel industry, when the stuff of sci-fi, has changed from speculative dreaming to a substantial reality. Organizations like Blue Beginning, established by Amazon's Jeff Bezos, and Virgin Cosmic, drove by Sir Richard Branson, are at the front line of this industry, creating suborbital spaceflight encounters for private people. The coming of reusable rocket innovation, exemplified by the upward departure and landing capacities of these shuttle, plays had a crucial impact in making space the travel industry financially possible.

The appeal of room the travel industry lies in the commitment of bearing the cost of regular citizens the amazing chance to encounter weightlessness and witness the ebb and flow of Earth against the background of the universe momentarily. Suborbital flights, albeit restricted in term, furnish members with a sample of the space explorer experience. While early space sightseers were ordinarily well-to-do people ready to pay over the top expenses for a short excursion past our air, progressing improvements propose a pattern towards more prominent openness and reasonableness.

The commercialization of room the travel industry stretches out past suborbital flights. Plans for orbital space lodgings, where visitors can dwell in microgravity for expanded periods, are in the calculated stages. These aggressive endeavors imagine a future where regular citizens, not simply prepared space travelers, can reside and work in space. The possibility of room lodgings brings up issues about the planned operations of long-term space home, life emotionally supportive networks, and the mental impacts of residing in restricted conditions past Earth.

The Worldwide Space Station (ISS), a cooperative exertion including various nations, has filled in as a microgravity lab for logical exploration and global collaboration. The possibility of business modules appended to the ISS, giving facilities to private people or exploration drives, mirrors the advancing scene of room use. As the ISS approaches the finish of its functional life, conversations about the progress to economically worked space natural surroundings pick up speed.

Colonizing other divine bodies, especially the Moon and Mars, addresses the following wilderness in human space investigation. The Moon, Earth's regular satellite, has recovered unmistakable quality as a venturing stone for additional space tries. Plans for lunar bases include the foundation of environments that can uphold human existence, direct logical exploration, and act as arranging guides for missions toward additional far off objections. The Moon, with its vicinity to Earth, offers a chance to test advances fundamental for supported human presence in space.

Mars, frequently viewed as the final location for human colonization, enamors researchers, designers, and space lovers the same. The difficulties of sending people to Mars and laying out feasible living spaces on the Red Planet are fantastic. Life emotionally supportive networks, radiation insurance, practical asset use, and the mental prosperity of space travelers during expanded missions are among the bunch factors that request cautious thought. Projects like SpaceX's Starship, intended for interplanetary travel, address a stage towards understanding the fantasy of a human presence on Mars.

The quest for extraterrestrial life, a longstanding logical undertaking, entwines with the story of room colonization. While Mars is an essential concentration because of its possibly livable circumstances previously, moons of the external planets, for example, Europa and Enceladus, hold captivating conceivable outcomes. These cold moons harbor subsurface seas that might actually have microbial life. The mission for indications of something going on under the surface past Earth stretches out our investigation to the furthest reaches of the planetary group.

Moral contemplations highlight the conversations around space the travel industry and colonization. The commercialization of room brings up issues about natural effects, space flotsam and jetsam the executives, and the evenhanded circulation of chances for space access. As confidential undertakings adventure into space, it becomes fundamental to lay out administrative structures that guarantee dependable space rehearses. The potential for asset extraction on heavenly bodies prompts moral consultations about saving the immaculate conditions of different universes.

In addition, the mental and humanistic parts of room colonization merit cautious assessment. The detachment and repression innate in lengthy span space missions, whether to the Moon, Mars, or past, present difficulties to the psychological prosperity of space explorers. The foundation of self-supporting states requires a more profound comprehension of human variation to extraterrestrial conditions. Besides, inquiries of administration and cultural designs in space states require smart thought to try not to imitate the weaknesses of Earth-bound frameworks.

The crossing point of room the travel industry and colonization with logical examination is a dynamic and harmonious relationship. While space the travel industry attempts add to the subsidizing and headway of room advances, they likewise act as stages for logical investigations. Microgravity tests led during suborbital and orbital flights offer experiences into major physical science, materials science, and natural cycles. The double reason nature of room the travel industry missions highlights the interconnectedness of investigation and logical disclosure.

Progressions in drive innovations assume a vital part in molding the potential for space the travel industry and colonization. The customary substance rockets that have moved space apparatus into space are developing to satisfy the needs of aggressive interplanetary missions.

Ideas like atomic warm impetus, which saddle the energy delivered by atomic responses for drive, could fundamentally decrease travel times for ran missions to objections like Mars. Productive drive frameworks are indispensable for arriving at far off divine bodies and laying out practical human presence.

Chasing space colonization, the use of in-situ assets turns into an essential objective. The Moon, for example, contains water ice in for all time shadowed locales, offering a likely asset for life backing and fuel creation. Mars, with its regolith containing minerals and ice, presents open doors for asset usage. Advances for separating and handling these assets are fundamental for making self-supporting living spaces past Earth.

The improvement of shut circle life emotionally supportive networks is one more basic part of room colonization. These frameworks mean to reuse and recover assets, limiting dependence on resupply from Earth. Developments in water reusing, air renewal, and waste handling add to the maintainability of long-length space missions. Shut circle frameworks are essential for colonization endeavors as well as for guaranteeing the strength of teams during broadened space travel.

In lined up with mechanical progressions, global joint effort assumes a focal part in the acknowledgment of room investigation objectives. The Artemis program, drove by NASA, imagines returning people to the lunar surface and laying out an economical presence. The cooperative viewpoint includes global accomplices, including the European Space Organization (ESA), the Canadian Space Office (CSA), and commitments from different nations. The Lunar Passage, a proposed space station in lunar circle, is one more cooperative undertaking that means to work with future lunar investigation and then some.

As we mull over the potential for space the travel industry and colonization, the thought of room as the normal legacy of all humanity comes to the cutting edge. The Space Arrangement, embraced by the Unified Countries in 1967, lays out standards for the tranquil utilization of space and denies the public apportionment of heavenly bodies. The moral structure of mindful space investigation stresses global collaboration, supportable practices, and the safeguarding of divine conditions for people in the future.

9.3 Reflection on the continued wonders and possibilities of rocketry in the cosmos

The adventure of rocketry, an odyssey that started with the red hot climb of the V-2 rocket during The Second Great War, has since developed into an entrancing excursion that rises above the limits of our home planet. As we ponder the proceeded with miracles and potential outcomes of rocketry in the universe, we leave on a heavenly investigation that traverses the domains of science, innovation, human undertaking, and the significant secrets of the universe.

Rocketry, at its pith, is a demonstration of humankind's dauntless soul of interest and investigation. From the early tests with explosive moved rockets in old China to the stupendous accomplishments of the Space Age, the direction of rocket improvement has been set apart by a steady quest for the unexplored world. The notable picture of the Saturn V rocket, conveying space explorers to the Moon during the Apollo missions, typifies the zenith of human inventiveness and assurance.

The cutting edge period of rocketry is portrayed by a different cluster of send off vehicles, drive frameworks, and mission targets. States, confidential ventures, and worldwide coordinated efforts add to a lively scene that stretches out from Earth's surface to the most distant compasses of the universe. The Worldwide Space Station (ISS), an image of worldwide participation in space investigation, circles above, filling in as a microgravity lab where logical tests and global organizations prosper.

Headways in impetus advances stand as a foundation of rocketry's development. Customary synthetic rockets, depending on the exothermic response among fuel and oxidizer, stay the essential method for arriving at circle. Be that as it may, advancements like particle drive, atomic warm impetus, and green fuels are reshaping the opportunities for interplanetary travel. These impetus frameworks hold the commitment of expanded effectiveness, decreased travel times, and broadened mission spans, introducing another period of investigation.

Reusability, a change in perspective spearheaded by organizations like SpaceX, addresses a weighty improvement in rocket innovation. The capacity to recuperate and repair rocket parts, particularly the main stages, in a general sense changes the financial scene of room travel. The Hawk 9's capacity to get back to Earth, land upward, and be ready for ensuing send-offs embodies an economical and savvy approach that denotes a takeoff from the disposable idea of early rocketry.

The wonder of room investigation stretches out past our nearby infinite area. Automated tests and telescopes adventure into the profundities of the nearby planet group and look upon far off systems, unwinding the secrets of the universe. The Mars meanderers, Soul, Opportunity, and Interest, cross the Martian surface, dissecting rocks, soil, and barometrical circumstances in the journey for indications of past or present life. The Hubble Space Telescope, roosted in low Earth circle, gives stunning pictures of far off systems, nebulae, and heavenly peculiarities, extending how we might interpret the universe.

Interplanetary missions, pushed by rockets that cross the huge region of room, epitomize the complexities and difficulties of grandiose investigation. The New Skylines mission, which led a memorable flyby of Pluto in 2015, uncovered a universe of frozen scenes and cryptic highlights. The Explorer tests, sent off in the last part of the 1970s, proceed with their odyssey past the limits of our planetary group, filling in as interstellar envoys conveying messages from mankind to the universe.

The potential for human colonization of other divine bodies poses a potential threat not too far off of rocketry's prospects. The Moon, with its closeness to Earth, remains as a characteristic objective for the foundation of lunar bases and examination stations. The Artemis program, drove by NASA, means to return people to the lunar surface, preparing for economical lunar investigation and filling in as a venturing stone for future missions to Mars and then some.

Mars, frequently named the "Red Planet," dazzles the human creative mind as an expected second home for mankind. The possibility of terraforming Mars, changing its air and environment to make it livable for people, addresses a fantastic vision that lies at the crossing point of sci-fi and logical hypothesis. While the difficulties of Martian colonization are considerable, progressions in drive, life emotionally supportive networks, and asset use bring this fantasy inside the domain of substantial chance.

The visionaries of the space business, from Elon Musk's SpaceX to Jeff Bezos' Blue Beginning, add to forming the direction of rocketry in the universe. Aggressive ventures like SpaceX's Starship, intended for interplanetary travel, and Blue Beginning's New Glenn rocket, pointed toward sending off payloads into space, highlight the confidential area's critical job in driving advancement and growing admittance to space. The serious scene of business space tries impels the business forward, encouraging an environment of quick mechanical turn of events and cost decrease.

The union of man-made reasoning (simulated intelligence) and rocketry proclaims another outskirts where AI calculations streamline mission arranging, route, and space apparatus activities. Man-made intelligence applications add to independent frameworks that improve the effectiveness and dependability of room missions. From the investigation of huge datasets to ongoing decision-production during space investigation, the combination of computer based intelligence with rocketry expands mankind's capacity to explore the intricacies of the universe.

Space the travel industry, an idea once bound to the domains of sci-fi, is currently turning into a reality. Suborbital spaceflights, presented by organizations like Virgin Cosmic and Blue Beginning, guarantee regular citizens a sample of weightlessness and a brief look at Earth from the edge of room. While space the travel industry is still in its outset, the possibility of regular people wandering past Earth's environment addresses a democratization of room access, welcoming a more extensive section of the populace to encounter the miracles of rocketry.

The ghost of room garbage, made out of old satellites, spent rocket stages, and pieces from crashes, represents a test to the manageability of room exercises. As the quantity of items in Earth's circle builds, the gamble of crashes and the age of extra flotsam and jetsam rise. Alleviating space trash requires worldwide collaboration, dependable space rehearses, and the improvement of methodologies for flotsam and jetsam expulsion and long haul space supportability.

Moral contemplations penetrate the talk on rocketry's proceeded with ponders. The natural effect of rocket dispatches, from the arrival of poisons to the consumption of the ozone layer, prompts reflections on the biological impression of room exercises. The moral utilization of space, directed by standards of global collaboration and the safeguarding of divine conditions, highlights the significance of dependable stewardship as humankind adventures further into the universe.

The reflection on rocketry's miracles stretches out to the potential for finding extraterrestrial life. The quest for livable exoplanets, directed by space telescopes like Kepler and TESS, expects to recognize universes where conditions for life might exist. The chance of identifying bio-signatures in the climates of far off exoplanets adds a layer of interest to the enormous investigation story. The disclosure of even microbial life past Earth would reshape how we might interpret the limitlessness and variety of the universe.

As we consider the proceeded with miracles and conceivable outcomes of rocketry in the universe, the human soul of investigation and the quest for information arise as directing powers. The accomplishments of the Space Age, from the main human strides on the Moon to the investigation of far off universes, act as guides enlightening the limitless capability of what lies past. Rocketry, as the vessel for our enormous excursion, drives us towards new wildernesses, empowering us to dream, find, and try the impossible.

The conceivable outcomes of rocketry in the universe unfurl as an embroidery woven with logical wonders, mechanical developments, and the getting through human mission for investigation. From the platforms of Earth to the farthest reaches of the universe, the direction of rocketry opens ways to conceivable outcomes that rise above the restrictions of our creative mind. As we dig into the huge breadth of grandiose potential, we experience a bunch of roads that shape the story of room investigation.

The groundwork of rocketry lies in the standards of drive, and the development of rocket motors has been a consistent excursion of refinement and development. Customary compound rockets, impelled by the burning of fuel and oxidizer, stay the workhorses of room investigation. Nonetheless, arising advances are growing the potential outcomes. Particle drive frameworks, controlled by electric fields speeding up particles, offer more prominent eco-friendliness and broadened mission abilities. As we peer into the future, atomic warm drive and other high level impetus ideas vow to alter interplanetary travel with more limited travel times and expanded payload limits.

Reusability remains as an extraordinary idea in the scene of rocketry. By and large, rockets were viewed as superfluous vehicles, disposed of after a solitary use. The approach of reusable rocket innovation, embodied by organizations like SpaceX with their Bird of prey 9 and Hawk Weighty rockets, has reshaped the financial matters of room investigation.

Recuperating and renovating rocket parts, particularly the primary stages, altogether diminishes the expense of sending off payloads into space, opening roads for additional incessant and financially savvy missions.

The potential for space the travel industry arises as a convincing possibility in the domain of rocketry. Organizations, for example, Virgin Cosmic and Blue Beginning plan to make suborbital spaceflights available to private people. Suborbital space the travel industry, described by brief excursions to the edge of room, offers regular citizens a sample of weightlessness and an all encompassing perspective on Earth against the background of the universe. While space the travel industry is at present in its outset, the thriving business holds guarantee for growing free to space and cultivating another time of investigation.

Interplanetary missions, pushed by strong rockets, set out on journeys of revelation to investigate the secrets of our nearby planet group and then some. Automated tests, similar to NASA's Diligence wanderer on Mars, cross outsider scenes, dissecting rocks and looking for indications of past or present life. Telescopes circling far off planets catch stunning pictures and unwind the insider facts of heavenly bodies lightyears away. As progressions proceed, the potential outcomes of uncovering extraterrestrial life, whether microbial or complex, become tempting possibilities that shape the eventual fate of room investigation.

The Moon, Earth's heavenly sidekick, arises as a venturing stone for future space tries. Lunar investigation, driven by both logical interest and key preparation, imagines the foundation of lunar bases and examination stations. The Moon's vicinity offers a proving ground for innovations fundamental for profound space missions, including life emotionally supportive networks, asset usage, and environment development. As we return to the lunar surface, the Moon turns into an essential stage for planning humankind for additional aggressive journeys to Mars and then some.

Mars, frequently hailed as the following wilderness for human investigation and expected colonization, catches the aggregate creative mind.

The Red Planet's interesting qualities, including a day-night cycle, polar ice covers, and the chance of fluid water underneath its surface, make it a convincing objective. Ideas of Martian bases, terraforming projects, and the foundation of self-supporting provinces coax another time of human presence past Earth. As we put our focus on Mars, the potential outcomes of turning into a multi-planetary animal varieties come into more keen concentration.

The visionaries of the space business, driven by nervy objectives and a promise to pushing the limits of plausibility, add to the unfurling story of rocketry. Organizations like SpaceX, established by Elon Musk, seek after aggressive tasks, for example, the Starship rocket intended for interplanetary travel. Blue Beginning, drove by Jeff Bezos, imagines a future where a large number of individuals reside and work in space.

The serious scene of private space undertakings encourages quick innovative turn of events and cost decrease, driving the business forward.

Man-made reasoning (simulated intelligence), with its capacity to dissect huge datasets and streamline complex frameworks, shapes a harmonious relationship with rocketry. AI calculations improve mission arranging, route, and rockct activities. Man-made intelligence applications add to independent frameworks that work on the effectiveness and dependability of room missions, permitting rocket to adjust to dynamic circumstances progressively. The combination of simulated intelligence and rocketry intensifies mankind's capacity to explore the intricacies of room investigation.

The investigation of space stretches out past our nearby planet group, with interstellar missions catching the creative mind of researchers and cosmologists. Ideas like Advancement Starshot imagine sending shuttle outfitted with smaller than normal tests to local star frameworks. Pushed by strong lasers, these tests could arrive at speeds moving toward a critical part of the speed of light, empowering quick investigation of far off exoplanets. The conceivable outcomes of wandering past our vast area challenge the limits of current innovation and spike conversations about humankind's position in the more extensive cosmic local area.

The test of room flotsam and jetsam, made out of old satellites, spent rocket stages, and different sections, highlights the significance of dependable space rehearses. As Earth's circle turns out to be progressively blocked, the gamble of impacts and the age of extra trash heighten. Relieving space flotsam and jetsam requires worldwide participation, the advancement of procedures for garbage expulsion, and the foundation of mindful rules for space exercises. Guaranteeing the drawn out maintainability of room attempts turns into a vital piece of the continuous story of rocketry.

Moral contemplations saturate the talk on the potential outcomes of rocketry in the universe. Natural effects, from the arrival of contaminations during dispatches to the exhaustion of the ozone layer, brief reflections on the biological impression of room exercises. The moral utilization of space, directed by standards of worldwide participation and the protection of heavenly conditions, highlights the significance of dependable stewardship as humankind adventures further into the universe. Adjusting the advantages of room investigation with the requirement for natural maintainability turns into a basic part of moral direction.

As we ponder the proceeded with marvels and potential outcomes of rocketry in the universe, the direction of investigation turns into an aggregate undertaking that rises above boundaries and ages. The amazing pictures of far off worlds, the many-sided dance of heavenly bodies, and the vigorous quest for information epitomize the embodiment of our inestimable excursion.

Rocketry, as the channel for human investigation and revelation, pushes us toward skylines that were once the area of dreams. The conceivable outcomes of rocketry in the universe welcome us to imagine a future where the marvels of the universe become a basic piece of the human experience.

www.ingramcontent.com/pod-product-compliance
Lightning Source LLC
LaVergne TN
LVHW041312200726
843509LV00009B/464